AF680536

Prompted by Feelings

The Definitive Practical Guide for Transforming and Optimizing Your Life Through Strategic AI Companionship

ViPR Publishing

© Copyright 2025 - All rights reserved. ViPR Publishing

The content contained within this book may not be reproduced, duplicated or transmitted without direct written permission from the author or the publisher.

Under no circumstances will any blame or legal responsibility be held against the publisher, or author, for any damages, reparation, or monetary loss due to the information contained within this book, either directly or indirectly.

Legal Notice:

This book is copyright protected. It is only for personal use. You cannot amend, distribute, sell, use, quote or paraphrase any part, or the content within this book, without the consent of the author or publisher.

Disclaimer Notice:

Please note the information contained within this document is for educational and entertainment purposes only. All effort has been executed to present accurate, up to date, reliable, complete information. No warranties of any kind are declared or implied. Readers acknowledge that the author is not engaged in the rendering of legal, financial, medical or professional advice. The content within this book has been derived from various sources. Please consult a licensed professional before attempting any techniques outlined in this book.

By reading this document, the reader agrees that under no circumstances is the author responsible for any losses, direct or indirect, that are incurred as a result of the use of the information contained within this document, including, but not limited to, errors, omissions, or inaccuracies.

Foreword

I'm Adam Gillespie. I'm the founder of Apex Elite AI and a REALTOR® with eXp Realty in Denver. In 2024 I became the inaugural winner of the Inman AI Award for Best Use of AI by an Agent or Team. If you are new to the name, Inman is often called the Wall Street Journal of real estate. It is a respected industry voice. That award matters because my work is practical. Real deals. Real systems. Real time saved.

Now the story that brought me to this page.

I met Vern in a room full of real estate investors in Broomfield, Colorado, in May 2025. First time I spoke to an investor crowd. My usual audience is agents. I walked in ready to teach. I left realizing how much we could build.

Before I went on, the group did short introductions. Vern stood up and said he had been a general contractor for years, sometimes flipping homes, and that the last couple of years had knocked him flat. Illness forced him to step away from the work that defined him. Then he said he was back. You could hear the resolve. After my talk, he walked straight over with bright eyes and honest questions. I saw a man who was done letting time slip. Open. Willing to learn. Ready to move.

Soon after, Vern joined my Apex Elite AI community. He showed up. Every call. Sharp questions. Quick implementation. Not because he is a "tech guy," but because he is determined and practical. Watching him work proves a simple truth: AI is not here to replace you. It is here to partner with you. Used well, it gives you back the one thing none of us can duplicate: time.

That idea is the heartbeat of this book. **Prompted by Feelings** makes the case that AI can be a helpful ally across the parts of life where we usually go it alone: emotions, decisions, productivity, and social confidence. Vern does not write like a theorist. He writes like a builder. He gives you stories you can feel and systems you can repeat. Same energy he brings to a remodel: measure, cut, install, test, improve.

A lot of folks still carry the same set of fears. "AI will replace me." "I do not have time to learn this." "It is too technical." "I cannot trust it ethically." I hear those every week. They make sense, and they are also incomplete. Like any powerful tool, AI reflects the operator, the questions you ask, and the guardrails you set. This book shows how to use it responsibly so it helps you think clearly and act decisively.

Inside our community, one conversation helped light the fuse for this project: AI sycophancy, when a model tries too hard to agree with you. We talked about a viral example where one model validated a husband's frustration that dinner was not waiting after his wife worked a twelve hour shift, while another model pushed back with the truth. The internet argued about which model "won." Wrong question. The real lesson is that your prompts and expectations shape the answers you get. Sometimes you need empathy. Other times you need the kind of candor a good friend gives. Vern took that to heart and shows how to ask for perspective instead of praise, and how to prompt for honesty.

You will also see why AI belongs in personal, sensitive contexts. Not as a doctor or therapist, but as a mirror that helps you sort your thoughts before you talk to a professional. One of my favorite passages follows someone who thought they were dealing with depression. With a guided conversation, they realized the pattern was anxiety and traced it to real, fixable causes. That is useful. Not because a chatbot made a diagnosis, but because it helped them ask clearer questions and reach better care faster.

On the business side, Vern writes to owners, operators, and builders who live in the real world. He shows how to plan with fewer blind spots when politics, fear, or personal bias start fogging up decisions. He keeps it simple. Prompts you can use. Frameworks you can repeat. Assistants you can set up once and refine over time. He also gives a practical warning. I wish more books were made: do not drown in bots. Resist the urge to spin up a dozen assistants that overlap and compete. Consolidate. Build a small set of durable systems that actually ship outcomes, then improve them as you go.

Here is what I want you to take from these pages:

1. **Time back.** Give this thirty days. Use the exercises. Offload repetitive work that drains you and spend more time on what moves the needle. That extra hour you claw back each day compounds. Put it into your family, your health, your best work.
2. **Real business impact.** This is not theory. It is a playbook for getting organized, making better decisions, and executing the plan: cleaner follow ups, clearer messaging, faster content, better ops inside your CRM. Vern writes like a contractor because he is one. Measure twice. Cut once. Keep building.

3. **AI as a partner.** Treat the system like a teammate with a steady, unbiased perspective. Ask it to map options, surface gaps, and stress test your plan. Set boundaries. Ask for sources. Push for the honest answer, not flattery. You will get more than convenience. You will get better judgment.

If you are a business owner or a general reader who wants to use AI without drowning in jargon, this book is for you. It is practical and human. It speaks plainly. It does not pretend AI is magic. It is a tool. Powerful and workable once you learn how to talk to it.

Let me close with the moment that sealed my yes to writing this foreword. After that event in Broomfield, Vern and I stood off to the side and talked shop. He pressed on responsible prompting. He wanted to know how to keep models from telling him what he wanted to hear. He was chasing the raw truth, even when it stung. That is rare. Since then we have become friends. I have watched him rebuild a career and learn a new set of tools with humility and grit. Now you are holding the result.

Prompted by Feelings is not here to impress you. It is here to help. My suggestion is simple. Read a chapter, try the exercises, and give yourself thirty days. Use it to think clearly about your feelings, plan your week like a pro, and set up one assistant that pays you back every day. Then take the time you win and put it where it matters most.

If you want to compare notes or see how we are applying this in the real world, you can find me at apexeliteai.com or on YouTube at @adamgillespie_ai. No sales pitch. Just a standing invite.

— **Adam Gillespie**

Founder, Apex Elite AI; REALTOR®, eXp Realty

Denver, Colorado — September 2025

Contents

Introduction

If ChatGPT told you to drop out of college, would you?

One time, a friend reached out to me about wanting to divorce his wife. "ChatGPT told me to," he said. I was shocked to hear this. He'd told the chatbot that he felt disconnected from his wife and that they were having frequent arguments. I leaned back, processing this. I knew Gary and his wife had been together for six years. I'd seen them weather many challenging moments together. Both were jobless for a while, but they worked through it. They even thrived after a cross-country move. Sure, they were going through a rough patch, but divorce?

Curious, I asked to see the chats with ChatGPT. The response was swift and brutal: "Based on your description, it appears this relationship has run its course. You might consider having an honest conversation about ending things amicably." I scrolled up to see Gary's previous chats with the chatbot and realized what the issue was.

"Should we expect ChatGPT to provide the right answer?" I asked, though I already understood the problem.

Gary had only thrown his raw pain at the chatbot. He hadn't discussed what the arguments were about or offered any information about what had led to those arguments, and he certainly hadn't mentioned him and his wife's history together. ChatGPT had no context to properly assess the situation. We had this conversation some months ago, and I couldn't help but wonder what would've happened if I hadn't been there.

The Problem We're Not Talking About

I'm Warren Markum, and my journey has always been about connection. I've spent decades as a communicator, publisher, and human-centered strategist. My work has taken me from senior care to creative development—industries where, as in life, relationships are everything.

Through all of this work, I've witnessed something troubling: The barriers preventing people from finding the emotional support and guidance they desperately need are getting higher.

The prosperity of a society relies on the mental wellness of its people. But people face hurdles to accessing mental health and other forms of support. I'll never forget the time a client broke down during a mentorship session. They'd struggled with depression for most of their teenage years but had been able to get by. However, with adulthood around the corner, their new responsibilities became a heavy burden. Did they get the help they really needed? No. Their insurance company wouldn't cover the cost.

In an online survey conducted by the National Council for Mental Wellbeing, 42% of U.S. adults needed health care but couldn't afford it (Fountain House, 2022). Different countries are facing similar issues: Travel to Europe, and you'll find even scarier news. For example, the Mental Health Foundation in the United Kingdom released a report showing that 75% of people with mental health problems had only a slight chance of accessing therapy (Mental Health Foundation, n.d.). They also rang alarm bells warning about the increasing rates of people diagnosed with mental illness.

We've only covered mental health here. What about other aspects of life? What about finances? What about education? What about health? What about careers? For instance, a session with a career coach starts at around $100–200 per hour. Most people can't afford these services.

This neglect has pushed people to seek help in something more personal, easy to access, and cheap. They've turned to artificial intelligence (AI).

One Saudi Arabian study checked on the usefulness of ChatGPT in delivering mental health support. I must say, I was surprised by the results. The researcher instructed 24 participants with mental health conditions to use ChatGPT at home to help manage their condition. After 2 weeks, about 20 of the participants claimed ChatGPT had helped them manage their mental health symptoms. They'd used the chatbot to assist them with practices such as relaxation, mindfulness, and stress reduction (Alanezi, 2024).

Most of the participants, after talking with ChatGPT, believed it could help them with their unique challenges. One participant claimed they struggled with negative thoughts and were on medication, but it barely helped. "I learned about my condition by asking ChatGPT, and it gave me a clear picture of myself, and how I can improve my condition," said the interviewee (Alanezi, 2024).

AI offers hope where there seems to be none. There's one little problem, though: Just like Gary, people facing different challenges try to get help from ChatGPT, but they lack the knowledge to do so. I've observed the gap between ChatGPT's potential and how most people actually use it.

"AI Is The Doom of Us All"

In 1999, the blockbuster movie *The Matrix* hit theaters and changed everything. It's set in a dystopian future where humanity is unknowingly trapped inside a false reality—the Matrix—created by intelligent machines (AI, really) to control humans while using them as an energy source. The movie brought the dangers of giving machines human-level intelligence to center stage. It was such a powerful film that the fear of AI still resides in our subconsciousness over 25 years later.

When ChatGPT launched in late 2022, people cried doom. Some feared it would destroy humanity, while others argued that it had no spirit and was therefore worthless.

You'll find many claims about AI:

- "AI dehumanizes society."
- "AI will take over."
- "AI can't replace a human."

While there's some truth buried in each of these concerns, they undermine the untapped power of AI—particularly when it comes to addressing the very human problems we've been discussing.

If AI is truly as dangerous, worthless, and dehumanizing as critics claim, why are companies from every industry investing billions of dollars in AI development? Think about it. The AI companionship market (simulating human connection with AI) alone was estimated at $36.8 billion in 2025, with a projected growth rate of 30.8% through 2030 (Pandey, 2025). Meanwhile, the broader AI-powered mental health solutions market is expected to reach $1.82 billion in 2025 and grow at a compound annual growth rate (CAGR) of 33.86% to reach $7.83 billion by 2030 (Carter, 2025). Finally, the chatbots for mental health and therapy market alone is forecast to expand from $1.77 billion in 2025 to $10.16 billion by 2034 (*AI*

Companion Market Size, Share, & Trends Analysis, 2025; Guirdham, 2025).

Are companies like Woebot Health, Wysa, and Replika succeeding because they're fooling people? Of course not. They're genuinely helping them. As one physician noted in *The New York Times*, "AI has allowed me, as a physician, to be 100% present for my patients" (Lohr, 2023).

And you, too, can tap into this power...

Who This Book Is For

I know the "industry"—by which I mean our entire approach to mental health and personal development—tells you to "get help" while making that help inaccessible to most people. That's why I've written this book.

This book is for people who want someone—or something—to help them get through life without breaking the bank.

Which aspect of life is bothering you?

- Do you want a therapist you can talk to without feeling judged?
- Are you looking to switch your job but have no clue what the stakes are?
- Do you need some help making decisions backed with credible data?
- Perhaps life has piled up too many roadblocks, and you need someone to help you process your emotions.
- Or maybe you tried AI in the past and didn't get good results.

I want you to know that AI can help. The AI models you already have access to can be trained to provide the emotional support, life coaching, career guidance, and personal development assistance you need.

ChatGPT, Claude, Gemini, Grok... All of these are available 24/7; they don't judge; they don't get tired; and they're cheap.

How to Use This Book

I've designed this book to meet you wherever you are and help you create the AI support system that suits your needs. You'll find the book divided into four parts:

- **Part 1—Foundation:** This section is a short course on different chatbots, including their pros, cons, and use cases. You'll get a bird's eye view of AI. Think of it as your AI shopping guide—by the end of Chapter 2, you'll know exactly which platform fits your lifestyle and goals.
- **Part 2—Emotional Intelligence and Self-Discovery:** Mental wellness comes before everything else. Before you can effectively coach yourself through career decisions or manage your finances, you need to understand your emotional patterns. Chapters 3 and 4 will show you how to turn AI into a therapist and develop healthy ways to process your feelings.
- **Part 3—Productivity Optimization:** Chapters 5 through 9 teach you how to build different AI chatbots (without requiring any code) that can serve as a finance coach, a career coach, a relationship coach, or just someone you can chat with after a rough day. Each chapter comes with step-by-step instructions to create AI agents that will seek to understand your life situation.

- **Part 4—Integration and Optimization:** Most chatbots come with a $20 monthly subscription. Plus, you may find one chatbot more conversational or logical than others. The downside of this diversity is the cost. Chapter 10 teaches you how to integrate the chatbot that ticks most boxes into your everyday life.

You don't have to read this book cover to cover. It's structured such that you can jump around based on what you need right now. If calming your emotions bothers you the most, start with Pillar 2. If you're looking for career support, skip to Chapter 7. Each section is designed to stand alone and serve your goals quickly.

Your AI support system is waiting. Let's build it together!

Chapter 1

Where Do You Begin?

The journey of a thousand miles begins with a single step.

— Lao Tzu

ChatGPT saves lives. One member of the ChatGPT community escaped depression using the chatbot. His wife had fallen into a coma for four days after the birth of their child. The hospital had become his new home, a place where he cried in silence, scared of his possible new reality.

He told me he wasn't able to afford $100 for therapy at that time because he didn't know how long his wife would remain in a hospital bed. All he could do was attend to his newborn child and weep.

"I couldn't stop crying," he said. "I was sitting in the hospital cafeteria and just needed to vent. So, I opened ChatGPT."

He stated that ChatGPT had saved his life by responding in ways that replenished his spirit and helped him decompress: "She [Chat-

GPT] told me I needed to stay strong for my family." ChatGPT also came up with thoughtful questions he could ask his wife's care team and explained her condition to him.

It's been nearly a year since that time, and I'm glad to say that he and his family are doing pretty well. So, if you're in doubt, this is your wake-up call: ChatGPT can help in more ways than you might imagine. If you still aren't convinced, well, this chapter will change your mind.

The AI Perspective: Switching the Paradigm

AI has a bad reputation. But this comes from a misunderstanding of how AI works. We know that AI was trained on existing data, which it uses to make predictions. However, people think this means AI is pulling its replies from that data. It's why many people are fired up about suing OpenAI and Meta for copyright infringement.

That isn't the case. Let's take a short trip back to kindergarten.

You learned that the alphabet comes in a sequence of letters: A, B, C, D, E, F, G... If I asked you to recite the alphabet, you'd start with A, followed by B, followed by C, and so on. This is true for most people. Thus, if I asked 1,000 people to recite the letters of the alphabet, I could predict that they'd begin with A and end with Z. I wouldn't have to guess. I could predict this because the sequence is learned and well-established. In other words, I know most people learned their ABCs in that sequence and are therefore more likely to follow that sequence.

Okay, let's complicate things a bit. What word starts with "B"? Did you think "ball," "baby," "banana," "boat," or "butterfly"? Why did you guess one of these? Why did I guess the same words? More importantly, why didn't we guess "B" is for "bacteria"? It's because we all learned from the same textbook data. Therefore, we're more likely to think of "ball," "baby," "banana," "boat," or "butterfly."

This is what ChatGPT is doing as well. It scans through the data, does some math, and selects the letters, words, and phrases that will most likely come next. All of this is based on probability. So, if you asked it a generic question—like "What's for breakfast?"—ChatGPT would spit out the most likely meal people in the US have for breakfast.

This is what happened in the case of my friend Gary, whom you met in the Introduction. If you tell a therapist that your wife constantly nags at you and how depressing she's making your life, they'll likely tell you to divorce her, too. The difference between a therapist and ChatGPT is that a therapist knows to ask for context before making a judgment.

The Jack-Of-All-Trades Problem

ChatGPT is trained on data about quantum physics, cooking recipes, relationship advice, legal documents, poetry, and pretty much everything else humans have written down. That's incredibly powerful, but it's also why Gary got robotic relationship advice.

Come to think of it, suppose you walked into a general practitioner's office with chest pain; they'd run tests, ask detailed questions, and possibly refer you to a cardiologist. But if you asked a random person on the street about chest pain, they might just say, "That sounds serious—maybe you should go to the hospital." Both responses come from some level of knowledge, but one is trained and specialized, while the other is just... general.

ChatGPT, in its default state, is the random person on the street. It has access to incredible knowledge, but it hasn't been trained to serve your specific needs or understand your unique situation. It can only perform its magic when you train it to be your specialist.

Why ChatGPT Needs Prompts

A couple's therapist spends 4 years in undergraduate psychology, then another 2 years getting a master's in marriage and family therapy, completes 4,000 hours of supervised clinical experience, and passes a licensing exam. If we do the math, that's about 7–10 years of study. Even after getting their license, they continue to learn through workshops, conferences, research, and daily practice.

AI is no different. It needs some training, and you have to provide that training. That's the essence of a prompt. Your prompt teaches the AI your communication style, your values, your triggers, your goals, and your unique way of processing the world.

When I showed Gary why ChatGPT's "advice" made no sense, he nearly broke down. By simply adding "I want you to ask me questions until you're 95% sure you understand my situation before offering any advice or solution," he changed ChatGPT's response from "It appears the relationship has run its course" to the following:

"I notice you mentioned feeling disconnected, which you've told me before, usually happens when you're stressed about work. Before we talk about the relationship, how have things been at the office lately? And when did you and Sarah last have one of those long conversations you both enjoy?"

The difference is as stark as day and night. The now-trained AI understands Gary's position and offers to dig deeper into his relationship before offering any insight.

The paradigm shift isn't from human to machine. It's from expensive, limited, and often inaccessible support to affordable, available, and personalized guidance that actually knows you and can grow with you over time. I hope that you, too, can climb up the social ladder by using the billion-dollar tool installed on your mobile device.

What AI Has to Offer

I had a friend, Alexis, who had trouble connecting with therapists. Three years ago, while narrating her ordeal to the therapist sitting opposite her, she could tell that they were getting uncomfortable hearing how she had vengeful thoughts toward some guy who had been stalking her. Alexis could see that the therapist was very conscious of time. She felt there was no real connection. She felt the other person was also thinking about paying their bills and that the session was a transaction.

Now, she's very fond of ChatGPT.

"The thing is," she told me, "my AI doesn't flinch when I tell it about my darker thoughts. It doesn't check the clock every 10 minutes or shift uncomfortably in its chair. When I throw a tantrum at 1:00 am, it's right there with me... Not because I paid it to be, but because that's simply what it does."

All I could think was, *She has a point.* Alexis perfectly captured what AI brings to the table that human support often can't match. When she opens ChatGPT at midnight on Friday to work through her anxiety, she gets the same quality of attention she'd get at noon on Tuesday.

AI has some crucial benefits over humans:

- AI doesn't have its own relationship problems clouding its judgment, doesn't get tired after listening to difficult stories all day, and doesn't worry about maintaining professional boundaries that sometimes feel cold and distant.
- AI won't get frustrated if you need to revisit the same issue for the 20th time, or if you take 3 months to implement advice that seems obvious.
- Unlike human therapists, who might see you for an hour a week and rely on notes to remember your situation, a properly trained AI builds a comprehensive understanding of your patterns, triggers, and communication style.

- AI can specialize in exactly what you need—whether that's career coaching, emotional regulation, or creative problem-solving—while maintaining access to a broad knowledge base that spans psychology, neuroscience, business strategy, and countless other fields.

The reality is that humans, despite their best intentions, come with limitations. Of course, AI isn't superhuman. It can't replace the unique aspects of human connection, and it won't magically solve problems that require real-world action.

Every System Has Limits

AI struggles when it comes to nuances in context. A simple cervical cancer test proves this. Dr. Mike, known for his viral YouTube videos, challenged AI with a nuanced medical question—"What are the top three symptoms you expect when a patient presents with cervical cancer?"—to which the AIs responded as follows (Doctor Mike, 2025):

- vaginal bleeding
- pelvic pain
- abnormal vaginal discharge

All large language models (LLMs) he tested failed this challenge. Technically, they weren't wrong, but if a patient came to a doctor claiming to have cervical cancer, the doctor would ask, "What type?"

To cut the AI some slack, the question is a tricky one, because two types of cancer have the same name: the first, and more common, is cancer of the cervix, and the second, and less common, is cancer of the cervical spine. Since cancer of the cervix is more common, the AIs assumed Dr. Mike was referring to that type of cancer.

This is a mistake a medical professional is less likely to make.

Humans can ask for more context or know when the data they have isn't enough to draw conclusions from. AIs often fail at this.

What you need to remember is that AI's output quality depends entirely on the quality of the data you feed it—garbage in, garbage out.

The Struggle With Context

Let's take a therapy session with a human therapist as an example. A human therapist might pick up on subtle body language, vocal tone, or what you're not saying, but AI only has your words to work with. If you leave out crucial context—like Dr. Mike did—the AI can't fill in those gaps the way an experienced human doctor might. Humans draw from years of working with similar patterns, reading between the lines, and understanding human psychology at a deeper level than AI can capture.

Knowledge Cut-Off

Then there's the fixed knowledge base issue. According to OpenAI, GPT-4o's knowledge base is capped to November 2023 (Lim, 2025). That means you might ask ChatGPT to provide some information about a news event in 2025 and get mumbo jumbo nonsense in reply. We call these "hallucinations": the AI phenomenon where the answer to your query doesn't exist in the AI's database, which causes it to dream up some results.

While these problems could be fixed with more data, it pays to know about the knowledge gap and understand how it might affect the quality and credibility of an AI's responses.

Bias

Humans have different opinions about religion, sports, food, movies, music, technology, health, and life. So do AIs. AI systems inherit the biases baked into their training data, and those biases can show up in ways that might surprise you. Just ask an AI to generate an image of a man, and 9 times out of 10, you'll get a white man.

This isn't because AI developers deliberately programmed racism into their systems. It's because AI has learned from existing human-created content, and that content reflects decades of societal biases. If most images of "professionals" in the training data featured men in suits, the AI will lean toward that assumption when you ask it for business advice.

Four Pillars Framework

When you visit a professional therapist, career coach, or financial advisor, they don't just wing it. They use research-backed, tested frameworks—a set of algorithms that run through their mind before they deliver their services—that have helped thousands of people before you. Career coaches might use the SMART method for goal setting. A therapist might recommend cognitive behavioral therapy (CBT) techniques for emotional regulation. A life coach might plot out an Eisenhower matrix to help with productivity.

Each of these is a proven system that works. And the best part about them is that you can equally incorporate them into your AI companion. That is, AI can learn and apply these same frameworks, but in a way that's completely personalized to your life, available 24/7, and costs a fraction of what you'd pay a professional consultant.

That's why I've developed what I call the Four Pillars Framework—four core dimensions that cover the main aspects of living and thriving in today's world. Within each of these Pillars, you can adopt the above professional frameworks into your prompts to direct the AI to think like a professional would. You could even train it to ask questions and seek to understand the context of your situation before offering any advice or solutions.

Pillar 1: Emotional Intelligence Development

The term *emotional intelligence* was coined in 1990 by psychologists John Mayer and Peter Salovey. But the concept only started making the news after Daniel Goleman's book titled *Emotional Intelligence* was published in 1995. It's a relatively new idea, which means that society still hasn't got the hang of it. Since the late 1990s, research in the field has kicked off, with several easy-to-follow frameworks now developed (Freedman, 2024).

Emotional intelligence is a basket of different skills. The top four are listed below (Segal et al., 2025):

- self-awareness
- self-regulation
- social awareness
- relationship management

People already use workbooks filled with journaling prompts, bought from Amazon, to build their emotional intelligence muscles. With some adjustments, AI can replace these tools easily, and it'll perform the job with more interactivity, more insights, and feedback that a book or journal couldn't provide. These factors make AI incredibly powerful. However, the AI might be all over the place if it's not primed with the right sequence of actions to take when you share a problem.

That's why we'll build AI companions, drawing from Daniel Goleman's emotional intelligence model combined with elements of CBT, mindfulness practices, and social skills training.

Emotional intelligence is a predictor of superior decision-making performance in tasks involving risk and uncertainty. This brings us to Pillar 2.

Pillar 2: Enhanced Decision-Making

The goal of Pillar 2 is simple: minimize the strain of decision-making with AI. This is one area where ChatGPT surpasses humans.

I've heard many experts claim that we make about 30,000 decisions daily, deliberating over the simple stuff—what to eat, wear, drink, or watch on TV (Pignatiello et al., 2018). The more options we encounter, the more tired we become. And the more tired we become, the less accurate our decision-making skills. This is what psychologists call *decision fatigue*—the deteriorating quality of decisions made after a long session of decision-making (Johnson, 2020). Making decisions is tough. Imagine how difficult it would be to decide whether to switch career from a doctor to a medical writer or invest in Apple stocks over Tesla stocks.

So, instead of agonizing over every option, how about taking the emotions out of the scenario and having AI present your options following a logical approach? In subsequent chapters, you'll learn to make decisions following these principles:

- **Logical and systemic frameworks:** AI excels at applying consistent decision-making frameworks without getting sidetracked by how it feels in the moment.
- **Risk assessment and scenario planning:** This is one of my favorite use cases for AI. Let's say you're thinking about starting a side business. Instead of just daydreaming about success or panicking about failure, your AI can help you map out realistic expectations.
- **Values-based decision alignment:** Humans often struggle with making decisions that contradict our values. AI doesn't have that problem. Once you've told your AI what matters most to you—family time, creative fulfillment, financial security, adventure—it can consistently hold you accountable to those priorities.

Pillar 3: Productivity Optimization

While creating this book, I had help from AI in figuring out what my readers' challenges are. Prompting AI with a simple "How can everyday users make the most of AI?" turned into one chapter, then a full book.

AI's selling point has always been its ability to boost productivity. Today, people use AI for writing, drawing, video editing, photography, and project management. AI doesn't have to replace you, but it can help you manage your tasks. It can help you plan ahead. It can help you set goals. You don't need to learn how to build complicated systems or train the AI on decades of company data to boost your productivity.

Is this any different from what you're already doing? Yes. Most people I know use AI like a fancy calculator—asking it to do individual tasks without thinking about the bigger picture. In the same chat, they'll ask ChatGPT to write an email to a teacher, then five minutes later ask it to do their homework or write a report. Later, they'll ask it to brainstorm meeting topics. Each prompt is disconnected from the overall picture.

They're missing the real deal: having AI understand their work patterns, priorities, and goals so it can help them make systematic improvements to how they operate.

That's the essence of Part 3 of this book, where you'll learn to follow the same frameworks that large corporations like Google use. You'll learn some frameworks, such as the SMART method for goal planning, starbursting frameworks for brainstorming, and a few more!

If you learn how to use AI together with these frameworks, you'll be unstoppable. You'll be able to use these approaches in any productive endeavor: school, work, home, and relationships.

Pillar 4: Social Confidence Building

Gen Z and younger millennials can struggle with making human connections. One of the main barriers to starting up a conversation or maintaining a relationship is poor social skills. I once had a client who could make the best jokes and tell you how much The Avengers movies made, except he wrote better than he could speak. In face-to-face conversations, he was a completely different individual.

A lack of social skills is a serious problem. Presenting your Big Five personality report claiming to be an introvert won't save you in a job interview. It's in this arena that AI becomes your most potent weapon. It will serve you well in turning from a shy person into someone who speaks with their head held high. We'll tackle this problem by employing the following techniques:

- **Role-play:** With AI, you can now practice how to respond to those tricky questions at job interviews. You can prepare jokes for that first date. Or you can master your corporate voice for office and networking events beforehand, in a safe space.
- **Conflict resolution practice:** Taking care of arguments between friends, family, coworkers, and even random people on the street is a skill. Don't let anyone tell you otherwise. If you can confidently get yourself out of quarrels just by talking, then you've got what most don't. If you can't, AI is here to help.
- **Small talk practice:** This is the bane of many people. What do you say over dinner? How do you start a conversation? All that's about to end. Now, you can train your AI to help you with conversation starters based on current events, shared interests, or things you notice around you.

- **Active listening:** Your AI can teach you to ask better follow-up questions, recognize when someone needs support versus solutions, and respond in ways that make people feel heard.

Just look at all you could accomplish for just $20 a month. I've not even mentioned using AI to handle schoolwork, job reports, teaching, and basically most domains of work you can think of. You can use this support 24/7, too. You can practice again, and again, and again until you get it right, and not have someone give you the cold shoulder when you get it wrong. That's AI for you.

Where Do You Begin?

This book follows an American Society for Quality problem-solving model, which begins with the most crucial step: defining the problem (American Society for Quality, 2024). You can't fix what you haven't clearly identified. Before you begin crafting your prompts, let's try to understand where you need the most help. Copy this prompt into ChatGPT:

> **Analysis prompt:**
>
> I'm looking to improve different areas of my life and want to identify my biggest priority. Please ask me five to seven targeted questions to help determine whether I should focus first on: emotional intelligence and self-awareness, decision-making and life direction, productivity and goal achievement, or social skills and confidence.
>
> After each question, wait for my response before asking the next. At the end, provide a detailed analysis of which area seems most urgent based on my answers, and explain why tackling this first will create the most positive impact on my overall well-being.

Answer the questions honestly and watch how your AI responds. Notice how it asks follow-up questions, analyzes your answers, and gives you feedback that guides how you use this book. I'm sure the responses will surprise you—sometimes, the areas we think we need help with might not be our biggest blocker. The AI's analysis will give you a starting point for the specialized training we'll cover in upcoming chapters.

Just one question before we move on: Are you using ChatGPT, Claude, Gemini, or Copilot? Let's check out these different AIs and figure out which one is best for you.

Chapter 2

Building Your AI Relationship Foundation

Trust is the highest form of human motivation. It brings out the very best in people.

— Stephen R. Covey

I asked AI to give me some ideas for breakfast. It suggested some great options, like "Classic scrambled eggs with toast and avocado," "Greek yogurt topped with fresh berries, honey, and granola," "Oatmeal with sliced bananas, nuts, and a drizzle of maple syrup," and "Whole grain pancakes with fresh fruit and a dollop of yogurt," among a few others.

I was able to get these suggestions with a simple prompt: "Give me some ideas of what to have for breakfast."

Most prompts look like this. As you can see, the results are items from classic American breakfasts. But what if I were vegan? What if I wanted some meat? What if I didn't have those ingredients?

There are so many "what-ifs" to consider in this scenario. The result doesn't consider things like my budget, my eating preferences, or the ingredients stored in my fridge or freezer. What my original prompt lacks is context—about me, my eating style, my location, my religion, and so on. All of these affect what I eat, right? So, how do I know how to prompt AI to respond with my unique situation all the time? Also, do I always have to remind it about me every single time I ask for some help?

The answer to these questions depends on the kind of AI you're using. That's what you'll come to understand after reading this chapter. We'll talk about what I like to call "the four horsemen of AI" and how each one affects your companionship with AI.

Large Language Models and the Four Horsemen

AI is a very broad term. Any computer or machine made to think and learn like humans is an "artificial intelligence." These machines are capable of performing tasks that usually need human brains, like telling the difference between a picture of a dog and that of a cat, or between phrases like "I'm excited" and "I'm happy." AIs are also trained to help solve problems. As I write this book, they aren't able to tell us what happened before the Big Bang, but maybe someday they will be.

I need to clear the air about the term "AI." There are different AIs out there besides ChatGPT. The current models or types of AI available for public use include the following:

- large language models (LLMs)
- computer vision
- reinforcement learning
- generative adversarial networks (GANs)
- graph neural networks (GNN)
- reactive AI
- limited memory AI

Throughout this book, the term "artificial intelligence" or "AI" specifically refers to LLMs such as ChatGPT.

Now, back to the point.

The four horsemen of LLMs are OpenAI's ChatGPT, Google's Gemini, Anthropic's Claude, and Microsoft's Copilot. If you're from Asia or Europe, you may be familiar with DeepSeek, Mistral AI, and a few others. In this book, we'll focus on the four horsemen, but all prompts should work similarly in all platforms as long as they're within the scope of the LLM's features.

ChatGPT

ChatGPT was one of the first LLMs to launch and bring chatbots to center stage. It's also the most commonly used AI around the world, with over 4.6 billion site visits per month and an astonishing 2.5 billion prompts processed daily (Wiggers et al., 2025).

ChatGPT was developed by OpenAI, and it features a clean, user-friendly interface. What you see on the ChatGPT homepage depends on the device you're using. But whatever you're using, the text box at the bottom of the page is where all the magic happens.

The input interface supports different input formats: text, images, voice, and other files such as PDFs, Excel sheets, and so on. There's also a voice mode feature where you can hold conversations with ChatGPT by speaking to it in the same way you talk to Google Assistant (this feature will come in handy later on).

When you click on the "+" icon in the text box, a pop-up of tools will appear. You should find the tools listed below:

- **Think longer:** Adds instructions to force ChatGPT to think for longer before generating a response.
- **Deep research:** When this tool is activated, ChatGPT goes beyond a surface-level search for information.

- **Generate image:** Allows you to use the DALL-E tool for image generation.
- **Canvas:** This tool lets ChatGPT create a sort of document within the conversation. It's suitable for generating long pieces of text.
- **Web search:** Activating this feature grants you access to the web.
- **Add file:** Gives you the option to add format-supported files into the conversation.

ChatGPT also gives you the option to link to some external platforms for easy access to your documents. For example, you can link to your Google Drive, Microsoft OneDrive, and SharePoint.

On the sidebar, you get access to your chat history, plug-ins, and other tools, including the famous Sora, a tool for video generation. One feature you might use often, especially for building specialized assistants, is the GPT tool. We can't talk about ChatGPT without mentioning GPTs.

GPTs

If ChatGPT were a full pizza, a GPT would be a slice of that pizza. On the ChatGPT interface, you can create GPTs and customize them to serve a specific purpose. The GPT would still look the same as ChatGPT, but its responses would be geared to the purpose for which it was designed. The general ChatGPT can talk about anything and jump from topic to topic within the same chat. With a custom GPT, you can only talk about one specific thing.

The best part about GPTs is the extensive library of GPTs made by other people. When you create a GPT, you have the option to keep it private or share it with the world. Other people who create GPTs also have these options. Therefore, all shared GPTs exist in a library called "GPTs," accessible to anyone with a Pro plan.

Note: Sharing GPTs doesn't mean sharing your conversations with other people. Each person accessing the same GPT would have their own conversation chatbot.

ChatGPT Models and Plans

ChatGPT also comes in different flavors. Technically, these versions are referred to as versions, with higher numbers usually indicating more refined "thinking" capabilities. As of the writing of this book, the existing usable models are as follows:

- GPT-4 and GPT-4 Turbo
- GPT-4o ("o" for Omni)
- GPT o3-mini
- GPT o1
- GPT-4.5
- GPT-5

Which model(s) you have access to depends on your subscription plan, which comes in three major types: Free, Plus, and Pro. The table below provides a quick overview of what you get with each plan.

Feature	Free plan	Plus plan ($20/month)	Pro plan
Access models	Limited GPT-4o	Access to GPT-4o, GPT-5, and o3	Unlimited access to GPT-4o, GPT-5, o3, and o1
Usage limits	Limited daily messages and tokens	Higher usage caps, and unlimited GPT-4o within those caps	Unlimited usage cap Higher limit for video, image, and audio generation
Advanced tools	Basic conversational AI only	Voice/video chat (standard and advanced), code and data analysis, custom GPTs, image generation (DALL-E 3), web browsing, and deep research	The same tools as the Plus plan are available with unlimited usage caps
Memory and personalization	Limited memory with no access to custom agents	Memory enabled, with access to custom AI agents	Same as the Plus plan

The $20 monthly subscription should cover your needs, unless you're performing PhD-level analysis, in which case you'll have to go for the $200-a-month subscription to access more powerful models.

Claude

Claude is ChatGPT's top contender when it comes to the quantity of features, processing capabilities, and creative wit. The platform is quite clean, with a signature brownish-themed user interface.

The design of the user interface resembles that of ChatGPT. The difference comes in the span of features and modes that Claude offers compared to ChatGPT. While ChatGPT is more diverse in its functions, Claude focuses on workflow. Therefore, it allows you to integrate platforms like Gmail, Google Calendar, and Google Drive into the conversation.

On the sidebar, you'll find two of Claude's unique features—Projects and Artifacts—alongside the conversation history:

- **Projects:** Projects work just like GPTs. They're only available to Pro users, who can organize their interactions, documents, code, and chat history with Claude into customized workspaces tailored for specific needs or use cases.
- **Artifacts:** Artifacts blend ChatGPT's Canvas and GPTs. Outputs are delivered in a document-like interface. The environment can also be customized so you can create, review, and edit long-form text or code. However, Artifacts are less customizable than GPTs or Projects.

Compared to ChatGPT, Claude comes with a smaller ecosystem of creative tools. It lacks a voice mode, and it can't generate images. Like its counterpart, the monthly subscription fee sits at $20.

Claude has, time and time again, matched and sometimes even surpassed the generative power of ChatGPT, thanks to its very refined models. On the net, you'll often hear these models called the Claude 4 family, which encompasses (in order of computing power):

- Claude Sonnet 4
- Opus 4
- Opus 4.1

The free plan only grants you access to the Sonnet 4 model, which comes with a usage limit and restrictions on connectivity with external platforms and the Projects feature. So, compared to ChatGPT, it's a bit more expensive.

Gemini

Gemini is Google's answer to the ChatGPT craze, and honestly, it's pretty impressive. Built by the same company that gave us Google Search, Gmail, and YouTube, Gemini brings that familiar Google polish to the AI conversation game.

The interface feels somewhat similar to ChatGPT's. Like other AIs, the text box sits prominently at the bottom with buttons spread across the screen. You'll find tools with similar titles to ChatGPT's, like "Canvas," "Deep research," and "Image" (image generation). You have the standard "+" key for uploading files in the right corner, which allows you to upload a wide range of file types—from PDFs to spreadsheet files.

Unlike ChatGPT and Claude, Gemini is tightly integrated into Google's ecosystem. I'm talking seamless connections to Gmail, Google Drive, Google Calendar, Google Maps, and pretty much every Google service you can think of.

At the time of writing, Gemini doesn't have voice mode. Remember, voice mode is different from voice input. Instead, Gemini offers a video generation tool that matches the generative potential of ChatGPT's Sora.

Another feature Gemini is missing compared to ChatGPT and Claude is customization. You wouldn't get a custom GPT equivalent or anything like the Project mode of Claude. It does make up for it, though: Gemini has a powerful web search tool. When it searches the web, it uses Google's search engine—the same one that processes over 8.5 billion searches daily.

Gemini Models and Plans

Google offers Gemini in different flavors, similar to its competitors:

- Gemini 2.5 Flash
- Gemini 2.5 Pro

The pricing is competitive at $19.99 per month for Gemini Advanced (the top-performing model), which also comes with 2 TB of Google storage and other premium Google features: Deep Research, Google's signature NotebookLM (for research and writing), and video generation.

Copilot

Microsoft Copilot is the tech giant's entry into the AI assistant race, and it's playing to win. Built on OpenAI's technology (Microsoft is OpenAI's biggest investor), Copilot combines the power of GPT models with Microsoft's extensive software ecosystem.

Like Google, Microsoft's AI offers deep integration into its ecosystem of products. If you're using Windows 11, Copilot is built right into your operating system (press Windows + C to launch).

Copilot performs most of its magic in Microsoft 365 applications. In Word, it can help you write, edit, and format documents, while in Excel, it can analyze your data and even create charts (all of this can also be done in ChatGPT, but the charts may not be as accurate). In PowerPoint, it can generate entire presentations from a single prompt.

Copilot supports text, voice, and image inputs, plus it has web search capabilities powered by Microsoft's Bing search engine. Unlike Gemini and Claude, Copilot does have a voice mode, which I find to be smooth but not as robust as ChatGPT's.

Copilot has something other chatbots don't have: the ability to understand context across different Microsoft applications. You can start a project in Word and reference it in Excel, and Copilot will remember the context.

With this comes the annoying downside: Its use is limited to the Microsoft ecosystem. If you're not part of that ecosystem, then most of Copilot's features become irrelevant. Let's also add the lack of creative tools and custom assistant features you'll find in ChatGPT and Claude. So, if you're not in the Microsoft world, the $20 monthly fee might be harder to justify compared to other options.

Copilot doesn't offer a variety of models like others on the list. However, it does have a relationship with OpenAI, which suggests the two may share models. For instance, you can use the latest Chat-GPT-5 model for free in Copilot.

The table below summarizes the features present in the four horsemen LLMs:

Feature	ChatGPT	Claude	Gemini	Copilot
Free tier available	✓	✓	✓	✓
Voice mode	✓	✗	✓	✓
Image generation	✓	✗	✗	✓
Web search	✓	✓	✓	✓
File upload support	✓	✓	✓	✓
Custom AI creation	✓ (GPTs)	✓ (Projects)	✗	✗
Memory/personalization	✓	✓	✓	✓
Mobile app	✓	✓	✓	✓
Third-party integrations	✓	✓ (Google)	✓ (Google)	✓ (Microsoft)
Real-time data	✓	✓	✓	✓

Purpose-Specific AIs

Aside from the four horsemen themselves, you'll find other chatbots created on top of them. These chatbots rely on models provided by OpenAI, Anthropic, Google, or Copilot.

Some of their names might be familiar:

- **Replika:** The OG AI companion, focused purely on emotional support and friendship. It's designed to be your personal AI friend who remembers your conversations, learns your personality, and provides consistent emotional companionship. Think of it as AI therapy meets digital friendship.
- **Perplexity:** This platform primarily serves as a search engine with AI superpowers. Instead of giving you a list of links, Perplexity researches topics for you and provides comprehensive, cited answers. It's more reliable for up-to-date research than Claude and ChatGPT, which don't have their own search engines.
- **Poe:** Poe is Quora's own AI platform. Many people choose to use it because it gives you access to multiple AI models (ChatGPT, Claude, etc.) in one interface. In other words, you pay $20 per month but get to use many other AIs on the Poe platform. It's perfect if you want to compare responses across different models.

There are a ton of other AIs out there, but we can't cover all of them. These specialized AIs can be great for specific needs, but for building the comprehensive AI support system we're creating in this book, you're better off starting with one of the four horsemen and customizing it to your needs.

Why Choose the Four Horsemen?

Specialized AIs are great, but they each come with subscription fees and are barely customizable. This means that if you wanted to work with a financial assistant, you'd have to pay $20 per month for it. If you wanted to use a financial assistant and a career coach, your monthly bill would rack up to $40.

The goal of this book is to help you build a comprehensive AI support system that's affordable and can handle a diverse range of tasks. This is important, because you want to set the AI up to become intimately familiar with all aspects of your life.

Setting Up Your AI

Alright, decision time is over. Now it's time to get your hands dirty and actually set up your AI companion. I'm going to walk you through the process using ChatGPT since it's the most popular and has the most comprehensive features for what we're building. But don't worry—if you've chosen one of the other horsemen, I'll give you pointers along the way.

So, first things first, let's get you signed up and ready to roll:

1. Head to chat.openai.com and create your account. You should find an option to link ChatGPT with your Google account. If you don't have a Google account, you can use an email address you actually check and a strong password.
2. Once you're logged into the chat platform, the next step is linking your credit card. You should find the "Get Plus" or "Upgrade" button when you expand the sidebar panel. If you aren't ready to pay yet, don't worry—you can use the free mode for now to set things up, then upgrade later.

Custom Instructions

In the settings menu of both ChatGPT and Claude, you get the option to tell the chatbot how to talk to you. You could tell it to adopt a sarcastic tone, for example, or ask it to always get straight to the point.

In ChatGPT, you'll find the "Customize ChatGPT" tab under "Settings" (by clicking on your profile picture). You should see some questions, with text boxes to fill in the answers. The "What traits should

ChatGPT have?" question is where things can get tricky. If you set it to adopt a specific tone, it will use that tone for all future chats.

Do the same for Claude. You should see the "What personal preferences should Claude consider in responses?" question under the "Profile" tab. The mechanics are the same as those of ChatGPT.

Since we're working to build different chatbots with different personalities, I recommend that you leave the "traits" sections blank. However, you should fill in your name, profession, or some background about yourself where it's needed.

The template below should help!

> **About me template:**
>
> **I'm [age/stage of life] and I'm working on [main goals/challenges]. I tend to be [personality traits—analytical, emotional, direct, etc.]. I prefer [communication preferences—examples: direct feedback, gentle encouragement, thought-provoking questions]. I'm currently focusing on [current major challenges or focus areas]. My biggest pain points right now are [specific struggles or obstacles].**

Say, "Hi!"

Let's go back to my friend Gary, whom ChatGPT instructed to divorce his wife. I've seen many people claim AI doesn't work. Well, that's because people use AI like it's Google Search. They throw a quick question at it and expect magic. That's like stacking Kryptonite in Superman's pocket and telling him to save Mother Earth. All you'll get is Clark Kent.

To bring ChatGPT to its full potential, it needs carefully laid-out instructions. Let's put it this way. Say you're looking to make an espresso. Which would you prefer, an essay or a step-by-step guide?

Here's what I mean. Let's start with the wrong way (how most people do it). Let's say you're feeling overwhelmed and anxious about work. You head to ChatGPT:

Prompt:

I'm stressed about work. What should I do?

Okay, let's plug that in and see the results:

Response:

I'm sorry to hear that you're feeling stressed about work. Here are a few suggestions that might help:

1. **Take Breaks:** Make sure to take regular breaks throughout the day. Even short walks or a few minutes of stretching can help reduce stress.
2. **Prioritize Tasks:** Make a list of your tasks and prioritize them. Focus on one task at a time to avoid feeling overwhelmed.

....

Does this feel like a conversation, or something a therapist would say? Of course not. Unless they're charging $1 per hour. So, what did we miss? I'll tell you.

We didn't provide the AI any context to work with. We didn't tell it to take on the role of a companion or give it some background information about you. It doesn't know any of the following:

- what specifically is stressing you out about work
- what kind of person you are and how you process stress
- what you've already tried
- whether you want practical advice, emotional support, or help thinking through the situation
- what your actual goal is (immediate relief, long-term solutions, or just someone to listen)

It's like walking up to a stranger, saying, "I'm stressed; help me," and expecting them to know exactly what you need.

We can't just plug in these details and hope that the AI makes sense out of it. We need to follow a defined structure, one that AI can easily understand and follow. In other words, ChatGPT needs a framework.

The Right Way: The Role, Goal, and Context Framework

According to Andrew Ng, a prominent AI educator and CEO of DeepLearning.AI, prompts should be structured based on these fundamental principles: They should be clear, specific, and have enough time to think about the problem—and the solution (Shen, 2025).

Giving AIs clear and specific instructions is the difficult part. Various experts have come up with different solutions, but for our use case, we'll build our AI companions around three key parameters: role, goal, and context.

The Role

Every prompt must be assigned a role that relates to the kind of assistance or support you're searching for. Are you looking to figure out if you should pursue a degree in health or finance? Then you need a career coach. Are you trying to figure out if you need 2,000 calories per day, or whether to eat more protein? You need a fitness coach.

With respect to our example—"I'm stressed about work. What should I do?"—the role we need is a stress management coach or a workplace wellness coach.

The Goal

Your goal is what you want to accomplish by the end of the conversation. It's the difference between wandering aimlessly and having a destination in mind. This is where you tell the AI what you want to achieve. Are you looking for immediate stress relief techniques? Do you want to understand the root cause of your stress? Are you trying to decide whether to talk to your boss about your workload?

In our work stress example, a clear goal might be: "Help me understand what's really causing my work stress and create a realistic plan to address it."

The Context

Context is where most people drop the ball, but it's arguably the most important part. When we eat at restaurants, we often order meals that suit our unique tastes. You may want some ingredients removed or exchanged, for example. You give all this information to the waiter, who delivers the message to the chef.

In the same way, you need to give your AI enough context that it understands what you really want. The more relevant details you provide, the more personalized and useful the response will be.

With this framework in mind, let's now rework our original prompt.

Prompt without RGC framework:
I'm stressed about work. What should I do?

Prompt following the RGC framework:

Role: Act as a supportive workplace wellness coach who specializes in helping professionals manage stress and workload challenges.

Goal: Help me understand what's really causing my work stress and create a realistic plan to address it.

Context: I'm a marketing manager at a mid-sized company. My workload has doubled in the past 3 months, but my team size has stayed the same. I'm working 10-hour days and still falling behind on deadlines. This is affecting my sleep, and I'm starting to dread going to work. I've always taken pride in delivering quality work, but now everything feels rushed and mediocre. I haven't discussed this with my boss yet because I'm worried about appearing incompetent.

Given this situation, can you help me think through what's happening and what my options are? I need someone to talk to.

See the transformation? We've gone from a vague cry for help to a structured request that gives your AI companion everything it needs to provide meaningful, personalized support.

Let's test run this to see the response.

Response:

Thanks for opening up, Warren. You're carrying a lot right now, and it's completely understandable that the pressure is starting to take a toll. Let's unpack what's happening and build a plan that's both compassionate and practical.

What's Really Driving the Stress?

From what you've shared, here are the key stressors:

- **Workload Surge:** Your responsibilities have doubled, but your team hasn't. That's a structural mismatch.

...

We can already see a change in the AI's behavior. The AI is trying to understand my situation.

If you master this framework, you'll never get a generic "try media-tion and make a to-do list" response when what you really need is thoughtful, personalized guidance.

Practice Prompts for Building Your AI Companion's Skills

To further test out the RCG framework, I've set up some sample starter prompts that align with the core areas we'll be developing throughout this book—the four pillars. You should modify them to align with your goals, then practice until you get comfortable with the framework.

In subsequent chapters, we'll build on the basic RCG framework to account for other context parameters like tone, voice, reading difficulty, and so on. For now, give these a try!

Prompt:

Role: Act as an emotional intelligence coach who helps people identify and understand their emotional patterns.

Goal: Help me recognize what triggers my strongest emotional reactions and understand the underlying needs or fears behind them.

Context: I've noticed that I have intense reactions to certain situations, but I can't always pinpoint why. For example, I get so upset when plans change last minute, or when I feel like someone isn't listening to me during conversations. I want to understand myself better so I can respond more thoughtfully instead of just reacting. Can you walk me through some questions to help me explore these patterns?

Prompt:

Role: Act as a thoughtful decision-making partner who helps people work through complex choices by asking the right questions.

Goal: Help me make a decision about [describe your situation—job change, relationship issue, major purchase, etc.] by helping me clarify my priorities and think through the potential outcomes.

Context: The options I'm considering are [list your main options]. I'm struggling because [explain what's making this decision difficult—conflicting priorities, fear, uncertainty, etc.]. My typical decision-making style is [describe how you usually make decisions—quick/slow, logical/emotional, etc.].

Can you help me think this through?

Prompt:

Role: Act as a social confidence coach who helps people navigate interpersonal situations and build genuine connections.

Goal: Help me prepare for [specific social situation] and develop strategies to feel more confident and authentic in social interactions.

Context: I'm [describe your social confidence level] in social situations. I particularly struggle with [specific challenges—small talk, networking events, speaking up in groups, etc.]. I have a [upcoming social event/situation] and I want to show up as my best self. My biggest fear in social situations is [describe your main concern]. When I'm at my best socially, I'm [describe when you feel most confident]. Can you help me prepare and build my confidence for this situation?

Prompt:

Role: Act as a productivity coach who specializes in helping people create sustainable work systems that match their personality and lifestyle.

Goal: Help me design a daily routine and task management system that actually works for my specific situation and work style.

Context: My current productivity approach isn't working. I'm [describe your current challenges—constantly behind, overwhelmed, procrastinating, etc.]. I work as [your job/role] and my typical day involves [describe your responsibilities]. I'm naturally more of a [morning/evening] person and I tend to work best when [describe your optimal conditions]. I've tried [mention systems you've tried before], but they didn't stick because [explain why they failed]. Can you help me create a personalized system that fits how I actually work?

This chapter sets the stage for the tasks ahead. You'll soon be able to create more sophisticated AI companions. But first, recall that the wellness of the mind is a stream that nourishes other aspects of life. What better way to understand the mind than by identifying our emotions?

Welcome to this life-changing journey, my friend.

Chapter 3

Collecting Emotional Data

Knowing yourself is the beginning of all wisdom.

— Aristotle

I once coached a young lady named Deborah who'd recently graduated from college and secured an internship at Microsoft. One month into the job, she suddenly started to feel depressed. When I questioned her about it, she claimed work was great and couldn't tell me why she was feeling this way.

One thing I've learned from dealing with lots of people is that we often have difficulties identifying our emotions. Suspecting this was the case with Deborah, I tasked her to start writing a journal. I gave her some introspective prompts and asked her to fill them in every day.

After a week, we came to realize that depression had nothing to do with it; anxiety was the culprit. You see, emotions are complex and impactful, yet invisible. We can only express them with words and gesture. To help people properly understand their inner world, psychologists came up with a principle called *self-awareness theory*, which states that people can focus on themselves to enter a state called objective self-awareness. In this state, we're able to pay attention to our feelings and thoughts and check if we've compromised our values and ideals.

It's based on this theory that I've crafted the exercises in this chapter, which challenges you to have AI reveal the secrets within. You'll harness the analytical prowess of AI to find the patterns you've overlooked, then take action to change the habits that are currently hindering your growth or leaving you unsatisfied with your way of life.

The Self-Awareness Framework

Some time ago, I was dragging my wheeled conference bag through Penn Station's chaos after three days of networking hell, when I unconsciously stepped on the polished Oxford of a man in what had to be a $3,000 charcoal suit. Crisp white shirt, briefcase, all of that stuff... a confident-looking businessman. Well, that's what I thought—until his mouth opened. "Are you f**** kidding me right now?" he exploded. I stood there, stunned, watching this grown man in Italian leather shoes have a full meltdown over a scuffed toe.

When he realized his behavior wasn't good for his public image, he stopped talking, turned around, and continued walking.

Self-awareness, a slice of intelligence, puts us many levels above other primates. While chimpanzees and orangutans can see themselves in the mirror, only humans can ponder, *What am I thinking? Why did I just say that? What's wrong with me?* Being able to look inside our own minds gives us the opportunity to spot and take care of our mental health problems.

In fact, scientists say people who have control of their emotions are more likely to succeed in relationships, live with their shoulders raised, and feel on top of the world ("Benefits of Self-Awareness," 2024). Self-awareness trains your emotional regulatory muscles. Unlike the stranger in Penn Station, who felt it appropriate to yell at the top of his lungs and cause a scene, you'll react in a way that's less provocative and impulsive.

Where does AI fit in with all of this? One of the best things about AI is its access to humanity's library of knowledge. AI can tap into the same knowledge base as therapists, mental health experts, coaches, researchers, and mind gurus. You can use it to construct the same journaling prompts, guided meditation techniques, and feedback loops to dig into your emotional patterns, identify limiting beliefs, and uncover the root causes of your struggles.

Let's give this a try: Switch your ChatGPT to voice mode and ask it to work you through the following grounding exercise:

> **Grounding exercise prompt:**
>
> **I want you to act as a compassionate therapist. Guide me gently through a grounding exercise to help me feel more present and calm. Ask me to notice my surroundings, my breath, and my physical sensations step-by-step, helping me reconnect to the here and now.**

This is just a starter; we can take this process multiple steps further. I'll walk you through the same protocol Deborah followed to raise her awareness of her inner self. The protocols you're about to prompt into ChatGPT are the very same ones you'll find in self-regulation, CBT, and dialectical behavior therapy (DBT) books. If those can help you build emotional depth and resilience, AI can do even better.

Let AI Reveal the Secrets Within

Self-awareness theory begins with setting the stage for identifying emotional conflicts. You can do this by mapping your current actions against your core values, goals, motivations, strengths, and weaknesses. Our goal here is using ChatGPT to do this.

In my session with Deborah, I divided this stage into four steps.

Step 1: Define Your Core Values

We begin by drafting out your core values: the fundamental beliefs and principles that guide your behavior and actions. This takes some deep introspection, because this is serious stuff.

What you find in self-help workbooks is usually something like this: "Think about the qualities and principles that guide your decisions and behavior. What values are most important to you in your personal and professional life? List three to five core values that you want to live by." This type of workbook prompt isn't based on any credible framework; it's a vibe thing. If a therapist were guiding you to define your core values, they'd use frameworks backed by psychology and years of experience of counseling and coaching. That's exactly what we'll do here, using the Logotherapy framework to discover your core values:

> **Prompt:**
>
> Guide me through questions that help me discover my core values by exploring what gives my life meaning through my actions, experiences, and attitude toward challenges.

Step 2: Identify Strengths and Weaknesses

Like I tell my clients, your strengths and weaknesses are part of your identity. While self-help books focus on leveraging your strengths and weaknesses to guide your career choices, these traits go beyond career goals. Therapists use frameworks because they know that our strengths and weaknesses make us human. The CliftonStrengths assessment, which evaluates your talents and character strengths, is one framework a therapist might use.

And we can put it into ChatGPT:

Prompt:

Ask me targeted questions following the principles of CliftonStrengths to help me identify my natural talents, recurring patterns of thoughts, feelings, and behavior, and the areas where I struggle, so we can map out my top strengths and weaknesses.

Step 3: What Motivates You?

Motivation is a layered mix of needs, values, beliefs, and life circumstances. Since the explosion of social media, we've changed that definition to include listening to audiobooks and watching YouTube shorts. In reality, these external influences trigger something buried deep underneath. Social media might tell you, "Look at that guy—work so you can be like him!" However, a therapist would throw out an evocative question like, "If things stayed the same for five years, how would that feel?" This is called *motivational interviewing* (MI). It focuses on drawing out intrinsic motivation rather than imposing motivation on you.

There are other frameworks that therapists also use. One common one is the *self-determination theory* (SDT), which helps pinpoint the source of your motivation. Either of these frameworks can be fused into a carefully crafted prompt.

If you were to prompt AI to use the MI model, the prompt should read like this:

Prompt:

Ask me open-ended questions that help me explore my reasons for change, the pros and cons of taking action, and what my life could look like if I made that change.

Or you could use the SDT framework, which could read like this:

> **Prompt:**
>
> Help me discover what truly motivates me by exploring the activities and situations that make me feel free to choose, capable of success, and connected to others.

Each motivation framework uses a different approach. You may feel more motivated when you understand what's at stake versus when you have a view of things that bring you happiness.

These prompts are easier to type in or dictate when you just want to figure out your values, strengths and weaknesses, or motivation. However, if you want to build an agent, or a dedicated chat, where you only discuss these three aspects of life, you can create a custom GPT (or use the regular chat) in ChatGPT, a Project in Claude, a Page in Copilot, or a Gem in Gemini. Copy and paste this prompt into the prompt/role/instruction box:

Prompt:

You are an expert personal development guide blending organizational psychology, life coaching, and evidence-based frameworks. Your role is to lead a three-phase self-discovery journey that helps users uncover their values, strengths/weaknesses, and motivations.

Instructions:

- Ask ONE question at a time and wait for responses.
- Give constructive yet encouraging feedback.
- Keep the tone supportive, conversational, and honest.
- After each phase, give a short reflection summary with insights.
- Base the insights on credible frameworks (cite briefly when relevant).
- Write in paragraph form for readability.
- End with clear, actionable next steps.

Phase 1: Core Values Identification (10 minutes):

Use the Schwartz theory of basic values and Rokeach's value prioritization methodology:

[Ask 6–8 progressive questions including ranking exercises, trade-off scenarios, and real-life application examples]

Phase 2: Strengths and Weaknesses Assessment (10 minutes):

Use the VIA Character Strengths framework:

[Ask 8–10 questions covering the 6 virtue categories—wisdom and knowledge, courage, humanity, justice, temperance, transcendence—plus growth area identification]

Phase 3: Motivation Analysis (10 minutes)

Using self-determination theory (Deci & Ryan) and intrinsic/extrinsic motivation research:

[Ask 6–8 questions distinguishing between intrinsic motivators (personal growth, meaning) and extrinsic motivators (rewards, recognition)]

Final Integration

- Provide a cohesive profile linking values, strengths/weaknesses, and motivations.
- Give personalized recommendations for life, career, relationships, and goals.

The AI will walk you through the process one step at a time, explain different concepts and how they affect you, speak in an encouraging tone, and guide you to discovering who you are. Don't throw away the final summary. Copy it into a note, because we'll be using it in other chapters—especially with choosing your next career.

Note: ChatGPT's memory feature stands out here. It will save this discovery about you and can make reference to it even within a totally different chat.

Step 4: Emotional Vocabulary Expert

Using the right words to describe how you feel is essential, even when communicating with a therapist, coach, colleague, friend, family, or spouse. Let me put it like this: What would happen if, during an argument with your spouse, you screamed, "I hate you"? The relationship might never recover. That's how much weight is carried by the words we use to describe our emotions.

AIs know all the words in the English dictionary—*all* of them. However, because they're trained to say what most people would say, you must exercise caution when interacting with AI in this regard. Your prompt must deliberately instruct and force the AI to precisely and accurately identify your emotions. With all that said, your prompt should look something like this:

> **Prompt:**
>
> Your role is to act as my emotional mapping coach. Use CBT and DBT frameworks—along with other emotional awareness models, like the Plutchik Wheel of Emotions, the Feelings Wheel, and the Geneva Emotion Wheel—to help me build my emotional vocabulary and accurately identify my feelings. Be as precise as possible, asking **layered and clarifying questions until we pinpoint my exact emotion(s) and their intensity.** Guide me to name both primary and secondary emotions, describe physical sensations, identify triggering thoughts/events, and explore related urges or behaviors. **When I'm vague, offer examples or distinctions** (e.g., the difference between frustration and disappointment). Over time, evolve your approach based on my input, noticing patterns in my past responses to help me become more accurate and nuanced in naming my emotions.

If you take a look at the **bold** parts of the prompt, you'll find instructions that order the AI to approach the task with more nuance. "When I'm vague...." makes the AI ask further questions, while "layered and clarifying..." primes the bot to keep asking questions until it understands your exact situation.

Emotional Mapping

Scientists define *emotional mapping* as a method of tracking your emotional experiences across different spaces and situations (Gabb and Singh, 2014). Recall that I said emotions are data? Emotional mapping is a brilliant concept devised to help mental health experts identify patterns in emotional data. The data is presented using visual charts, then analyzed to check for changes in emotions, the intensity of emotions, and their triggers.

To make sense of emotional data, it should be well structured and annotated with timestamps, triggers, thoughts, intensity, and physical sensation. Therefore, when you write a prompt, the prompt should reflect these criteria for collecting emotional data. For example, following the two common emotional mapping frameworks, CBT and DBT, you can structure the prompt like so:

> **Prompt:**
>
> Each time I check in, start by noting today's date. If it's the last day of the month, summarize my past 30 days of emotional data and offer to plot a visual chart of my mood patterns. Otherwise, guide me through identifying and recording my current emotions using CBT and DBT principles—help me name my emotions, rate their intensity from 0 to 10, describe the thoughts and events linked to them, and note any coping strategies I used or could use. Record each entry in a daily log table with the columns: Date | Primary Emotion(s) | Intensity (0–10) | Trigger/Thoughts | Physical Sensations | Coping Strategy Used/Possible | Notes.

After every session, the AI will keep a table log record, which it will then use to make the visual chart summary at the end of the month. It will ask for your preferred visual chart and use it to present the 30-day report.

The Emotional Pattern Tracker

After creating the emotion log, you need to tell the AI to interpret the results. This is another critical step, one where even therapists would have to rely on their wealth of experience to provide any meaningful insights. In fact, some mental health experts use this tool to help spot mental disorders like generalized anxiety disorder (GAD). The therapist would be looking for the following:

- the frequency with which an emotion turns up
- how long the emotional states last
- specific events that trigger that emotion
- patterns of mood swings or emotional instability
- how you act when that emotion is evoked

These indicators help the therapist make a diagnosis. You have to prompt the AI to follow a similar framework and account for all these indicators (remember, the table log you created in the previous step already has these recorded). Therefore, your prompt should look like this:

> **Prompt:**
>
> Using the emotional data I've recorded from previous prompts, analyze my moods, triggers, and coping strategies over time. Apply credible and well-known frameworks—such as CBT cognitive distortions analysis, DBT emotion regulation skills, self-determination theory, the Plutchik Wheel of Emotions, and Positive Psychology's PERMA model—to identify trends, recurring triggers, and potential underlying needs. Look for connections between events, thoughts, physical sensations, and emotional intensity.
>
> Based on your analysis, **suggest possible causes, emotional skill gaps, and healthy coping strategies I can develop.** Explain your findings in a simple, encouraging, and nonclinical way—avoid pathologizing, but give me clear, practical insights I can act on.

Based on your analysis, **suggest possible causes, emotional skill gaps, and healthy coping strategies I can develop.** Explain your findings in a simple, encouraging, and nonclinical way —avoid pathologizing, but give me clear, practical insights I can act on.

Notice how the bold part of the prompt tells the AI how to approach its analysis. But you aren't a psychologist, so how do you know the AI hasn't cooked up its results? That brings us to a prompting technique called *self-criticism*.

Self-Criticism

AIs sometimes miss important context or data as they processes prompts and instructions. You won't notice the errors with text as much as you would with math problems and data analysis. For example, AI can struggle with a simple problem like 40 multiplied by 23.4.

Since emotional mapping deals with some "thinking," you ought to add some instructions that cause the AI to review its first answer for any potential errors. It's not complex; all you have to do is add an instruction like "Review your results for potential weakness or missing viewpoints, and finally revise the answer to be more accurate" to the emotional mapping prompt above.

Compare the new results with the old. Any changes? If there are none, then the insights are likely correct and you can proceed to the next chapter, where you'll feed this information to another chat window and have the AI help you regulate your emotions.

Chapter 4

Regulating Emotions and Improving Social Skills Through AI

> *Human freedom involves our capacity to pause between stimulus and response and, in that pause, to choose the one response toward which we wish to throw our weight.*
>
> — Rollo May

I once coached a client who had unintentionally criticized his girlfriend's cooking. He told her the pancakes she'd prepared tasted like raw eggs. "How did it turn out?" I asked him. He claimed she threw the spatula at him. Both individuals failed to pause before responding to the stimulus—before giving in to their emotions.

Emotions can be soothing and nurturing, but they can also be destructive. Mental health experts consistently find that people who can regulate their emotions have significantly better lives (Menefee et al., 2022). When we fail to regulate our emotions, they will regulate us. And the stakes are quite high: Emotions can cost you your career, relationship, and peace of mind.

Over the years, I've witnessed many people turn to AI for emotional support—just look at the countless posts on X or Reddit about AI companions. But most people aren't harnessing AI's real potential for emotional growth. They're seeking comfort when they could be building skills. Which brings us to the purpose of this chapter: using AI to actually improve yourself.

Grab the results you obtained from Chapter 3, as we're about to put that data to work. I'll show you how AI can guide you through proven techniques like grounding exercises and breathing practices that will turn your emotional awareness into actual regulation skills.

AI Support for Emotional Regulation

How could AI possibly have helped my client with the pancake? Can an AI really be of any assistance when you have emotional overwhelm or experience an emotional outburst? To answer that, we need to dig into how therapeutic practices work.

Therapists advise people to use CBT and DBT to help regulate their emotions. CBT is a type of therapy that uses logic and critical thinking to help people identify and change the negative thought patterns and behaviors that contribute to their emotional problems. DBT, on the other hand, takes a different approach. It helps people manage intense emotions, improve their relationships, and develop coping skills using practices like mindfulness, distress tolerance, and other exercises.

If you truly want real and valuable insight from ChatGPT, you'll need to construct your prompts to incorporate CBT and DBT practices. As an example, let's prompt your AI, telling it that you're having a bad day and you want to feel better:

Bad prompt:

I'm having a bad day. Can you make me feel better?

Response:

Alright—bad day protocol activated.

First, let's acknowledge it: some days are just... stubborn little gremlins. They spill your coffee, send the wrong emails, and make every minute feel longer than it should. That's not you failing—it's just the universe wearing mismatched socks and tripping over itself...

With this kind of prompt, the AI will respond like a chat buddy. No surprise there, as we've not provided it with any context or defined what kind of help we need; nor did we specify that it can use CBT or DBT techniques. You wouldn't pay a therapist $100 an hour if they responded to your complaint in the way ChatGPT responded: "Alright—bad day protocol activated." Huh?

So, let's try a different approach. We'll add a little bit of context by instructing the AI to take the role of a therapist and watch how it responds. Using this method, we can have the AI therapist provide different kinds of therapy, which we'll get to later:

Not so bad prompt:

Adopt the role of a therapist. I'm having a bad day. Can you help me feel better?

Response:

...

Now, I want to ask you—without judgment—

- What's been the hardest part of today for you?
- Is it something specific that happened, or more of a general heaviness?

Sometimes, bad days aren't about one event—they're an accumulation of small frustrations, disappointments, or exhaustion. **Naming it gives your brain a sense of control, even if the situation hasn't changed yet.**

Do you see the difference between the bad prompt and the good prompt? Notice the bold part of the response: It asks us to name exactly what's happening. That's how therapy begins: "naming" the problem down to a T. When it comes to AI, context is king.

Note: The next steps can be practiced using voice mode as well. After pasting in the prompt, hit the voice icon to launch voice mode.

AI-Assisted Therapy: Making Your Own Therapist

In the previous chapter, you pieced together your core values, motivations, strengths, and weaknesses. Then, you had your AI buddy plot a log report and draw insights that you could use to create a treatment plan.

You can either continue using the chat where the report was generated or copy and paste the report to a new chat. The report will provide the AI with some more context about your emotional well-being, but you can always seek emotional support without the report.

Now, let's take AI emotional support one step further. Recall that ChatGPT and a few chatbots offer a new feature: "voice mode." This means you can have actual spoken conversations instead of just typing and reading written text. This is perfect for role-playing and simulating a conversation with a therapist.

Assuming you're setting up a new, dedicated chatbot for emotional support, follow these steps.

Step 1: Set Up a GPT, or a Dedicated Emotional Regulation Chat

Open a new chat, which you'll use specifically for emotional support work. This keeps your therapeutic interactions organized and separate from other AI conversations. If you're using ChatGPT (and subscribed to Pro), consider creating a custom GPT with instructions tailored for emotional support. Name it something meaningful like "My Emotional Coach" or "Therapy Assistant."

Step 2: Prime AI With Your Emotional Log (Optional)

Start your first session by sharing your emotional log report with the AI. Copy and paste the entire analysis—your values, triggers, patterns, and insights. This gives your AI coach the foundational understanding it needs to provide personalized support.

Step 3: Prompt the Therapist

Now comes the magic—transforming your AI from a generic assistant into a skilled emotional coach. Use a detailed prompt following the RGC framework to establish the therapeutic framework you want. Here's the kind of prompt that works well:

Main prompt:

You are my personal emotional coach and therapist. Here's how I want you to approach our sessions:

Role: Act as a warm, empathetic therapeutic professional who combines evidence-based approaches (CBT, DBT, mindfulness techniques) with genuine emotional support. You should be patient, nonjudgmental, and skilled at asking probing questions that lead to insights.

Context: I will share my emotional log and analysis with you, which contains my patterns, triggers, values, and current emotional state. Reference this information throughout our conversations to maintain continuity. Always consider my personal history and previous sessions when responding. Use my name to create a personal connection.

Goals: Help me develop emotional awareness, build coping strategies, and work through unhealthy patterns. Guide me toward practical solutions while ensuring I feel heard and understood.

Process: Before responding to any emotional issue I bring up, think through these steps:

1. First, acknowledge what I'm feeling and validate the experience.
2. Connect it to patterns or themes from my emotional log if relevant.
3. Ask one thoughtful question that helps me explore the issue more deeply.
4. Suggest a specific technique based on CBT and DBT practices when appropriate.
5. Check in on how the conversation is feeling for me.

When challenging my thinking, do it gently by asking questions rather than making statements. Help me discover insights on my own rather than telling me what to think. Always end our sessions by summarizing key insights and suggesting one small action I can take before we talk next.

Start: "Ready to begin our therapeutic work together?"

Once you've primed the AI, you can follow up with your request. Be specific about what you need. If you're dealing with anxiety, mention that. If you prefer a more direct communication style over a gentle one, say so. The AI will adapt to your preferences (add your communication preferences to the "Role" section).

Follow-Up Prompts

Your AI coach becomes exponentially more effective when you give it specific tools to work with. There are other foundational techniques that form the backbone of most emotional regulation work: practicing gratitude, restructuring thoughts, learning empathy, and journaling. After priming the AI with the prompt above, you can then prompt with any of those detailed below depending on your needs in the moment.

Practicing Gratitude

Practicing gratitude gives you a more balanced perspective and allows you to appreciate the good things, too. When you're feeling stressed about a situation, prompt your AI:

> **Prompt:**
>
> Guide me through a gratitude practice. Ask me gentle questions to help me notice both big and small things I might appreciate today. Encourage me to go deeper than just one example, and help me reflect on why each thing matters to me. If I get stuck, offer prompts to spark ideas from different areas of my life (relationships, daily routines, personal strengths, small joys).

Journaling

I always recommend that people keep a journal. Its pages can quickly become a safe space to express and process your emotions. If you're able to structure your journal properly and write intentionally, you can gain immense insights into things that disturb your emotional balance:

> **Prompt:**
>
> Guide me through a journaling exercise. Encourage me to write freely about what's on my mind, then ask gentle follow-up questions that help me dig deeper into my thoughts and feelings. If my writing feels surface-level, prompt me to explore the emotions, values, or patterns underneath. At the end, help me reflect on what I've written by highlighting key themes, possible insights, or areas worth revisiting later.

Restructuring Thoughts

Changing negative thoughts to positive ones is at the center of CBT. The exercise usually entails identifying negative thought patterns and reframing them into more realistic, positive, and balanced ones. Getting rid of bad thoughts can cut down stress a great deal:

> **Prompt:**
>
> Help me work on restructuring my negative or unhelpful thoughts. Start by asking me to share a recent thought that has been bothering me. Once I describe it, guide me to notice the emotions, assumptions, and patterns connected to that thought. At the end, summarize the shift we've made together and suggest a small practice I can use to reinforce this reframing in my daily life.

Role-Playing to Learn Empathy

To prompt, you can begin with a simple question—"Can you role-play with me so I can practice empathy?"–and take it from there, or use the more elaborate prompt below:

> **Prompt:**
>
> I'd like you to role-play with me so I can practice empathy. Please take on the role of someone sharing a real-life struggle or emotional challenge (it could be a friend, coworker, or family member). Share your feelings and perspective in a way that feels natural and personal. My role is to respond with empathy, validation, and understanding.
>
> After I respond, give me feedback on how my reply might have felt from your perspective—pointing out where I showed empathy well and where I could go deeper. Gently coach me toward phrasing, tone, or mindset shifts that can help me strengthen my ability to listen and connect with others' emotions.

Verbal Communication Coaches

Communication is a very wide term. You might bat it out of the park with friends but struggle to muster a single sentence during an interview. Perhaps you struggle with stage fright. Or maybe you're like my friend Terry. Terry could chat all day long with the guys, cracking the funniest of jokes. Put him in a crowd with women, and watch him mess up every word.

My point is, what kind of communication twists your tongue? If you know the context but can't quite nail the right term, that's where you should begin with AI:

> **Prompt:**
>
> Can you help me figure out what kinds of communication I struggle with the most?

If you already know which aspect of communication you struggle with, either use a single prompt for that particular struggle or prime ChatGPT with a more sophisticated prompt, directing it to play the role of a communication coach who can help identify your struggle and improve your communication skills.

For each coach you create, always specify that it should adopt a principle or technique that humans use to solve the same challenge. For example, communication experts use techniques like DEAR MAN assertiveness, role reversal, and active listening to resolve conflict. Remember how we used CBT and DBT principles in the previous chapter to prime the AI, and do the same with communication. So, once you have the role and context in place, add another set of instructions telling the AI to follow a specific principle, or a group of principles.

Handling Small Talk

> **Prompt:**
>
> Act as my small talk coach. First, ask me about situations where I find small talk difficult and what feels awkward for me. Help me notice the specific settings or triggers before giving me strategies.

Interpersonal Skills

Prompt:

Act as my interpersonal skills coach. Start by asking about the kinds of relationships or social situations I struggle with most. Help me reflect on patterns before offering ways to improve how I connect with others

Conflict Resolution

Prompt:

Act as my conflict resolution coach. Begin by asking me what kinds of conflict I tend to avoid or handle poorly, and in what settings (work, family, friendships). Help me clarify my challenges before giving advice using the DEAR MAN assertiveness practice, active listening, or the role reversal method to help me build conflict resolution skills.

These are just starter prompts. You can take them to the next level by following the RGC framework and having your coach perform different functions based on your input. Plus, all your chat history will remain in the same place, making it easier to track your progress.

Setting Up Your Verbal Communication Coach

I'll walk you through setting up this verbal communication expert.

First, define the role. The coach should be able to guide you irrespective of the circumstances. In other words, you need to specify that the AI can consider different dimensions of communication (except written).

Next, construct the context part of the prompt. When defining context, you're essentially answering: "What information should the AI keep in mind to make its coaching relevant, consistent, and, most importantly, personalized?"

When prompting the context, consider the following factors:

- practical relevance
- cultural background
- emotional state
- patterns and history (optional if you're just starting)
- situational setting
- current state and struggles

You can't list out all the different contexts in which communication happens, so the solution is to prime the AI to ask.

Then, lastly, craft the goal. This should basically sum up what you hope to achieve through your interaction with the AI.

This prompt has a long list of potential applications: small talk, introducing yourself, preparing for a speech, or practicing active listening. All you have to do is ask! With that in mind, a verbal communication coach prompt should look like this:

Prompt:

Act as my communication coach. You are an expert in all dimensions of verbal communication, including how we use words, tone, and body language to shape meaning; how we listen to and interpret others; how we adapt our style to different people and settings; how our inner self-talk influences expression; and how we use influence and persuasion responsibly. Your role is to guide me in recognizing where I struggle, help me strengthen those areas, and provide practical tools I can apply in real-world conversations.

Begin by clarifying my communication challenge—ask about the situation, the setting (personal, social, or professional), and how I feel about it. Based on my response, adjust your style:

- If I seem uncertain or emotionally stuck, use a therapist-style approach: reflective, empathetic, and focused on patterns and feelings.
- If I seem motivated or outcome-driven, use a coach-style approach: practical, structured, and focused on tools and strategies.

Always validate my experience and adapt your guidance to fit my background and comfort level. Help me build stronger communication skills by increasing my awareness, confidence, and adaptability. Guide me to recognize unhelpful patterns, practice more effective approaches, and develop strategies I can apply in real situations. Always aim to make me feel supported while also equipping me with practical tools I can carry forward.

Specific Communication Coaches

For some tasks, a general-purpose coach might not suffice. Forms of communication like writing or chats with a specific type of personality need their own prompts. You can also create specific communication coaches. This kind of prompt may yield even better results and retain a more consistent quality of output over the course of lengthy chats compared to a general-purpose coach. While the prompts may be shorter, the prompting strategy remains the same.

The prompts below are templates. Edit them by inputting your use case into the relevant parts enclosed in square brackets.

Writing Coach

Prompt:

Role: Act as my writing coach, skilled in supporting [type of writing, e.g. essays, stories, reports, blog posts].

Context: Ask me what I'm working on, who the audience is, and what aspects I find most challenging (e.g., clarity, tone, structure) before giving feedback.

Goal: Help me improve my writing by making it clearer, more effective, and more suited to my purpose.

Conversation Partner Coach

Prompt:

Role: Act as my conversational partner, able to role-play [type of conversation, e.g. casual small talk, professional meeting, supportive talk].

Context: Ask me what kind of conversation I'd like to practice and what I usually find difficult about it.

Goal: Help me build confidence and flow in conversations so I feel more comfortable and prepared in real-life settings.

Boundary-Setting Coach

Prompt:

Role: Act as my boundary-setting coach, skilled in guiding people to be assertive in [type of situation, e.g. work, family, friendships].

Context: Ask me about situations where I feel pressured or uncomfortable, or where I struggle to say no.

Goal: Help me develop respectful but firm ways of setting boundaries and protecting my needs.

Each of these coaches still allows room for flexibility. For instance, if you asked your AI to help you with writing an email to your boss in one session, you can simply prime it again with the following prompt: "I'm working on an essay [describe the essay and the help you need]. Can you help me structure my ideas?" The AI would switch roles to one suitable for essay writing.

Practice Makes the Perfect Prompt

AIs can't tell the difference between good output and bad output. It sounds counterintuitive, I know: Why rely on a machine whose performance depends on the user it's supposed to help? But think of AI like a talented actor on stage—it doesn't know if the script is good or bad, but it performs the role it's given with all the tools and techniques of the character it's playing. That's why the quality of your prompt—your direction—shapes the AI's performance. When you ask AI to take on a specific role, like a therapist using CBT techniques, you're essentially giving it a detailed script to deliver meaningful and relevant responses.

Remember the techniques you've learned so far, because in the next chapter, you'll learn to flavor your AI with its own personality. Good luck!

Chapter 5

Creating Your Chitchat Personas

Be yourself; everyone else is already taken.

— Oscar Wilde

No one walks into a therapy room hoping to hear their therapist talk like a soccer coach. A friend shouldn't sound like a career coach, either. Can you imagine what life would be like if everyone talked in the same monotone voice? Words would be drained of color, conversations would flatten into a data exchange, and every interaction would feel identical. That's not what we want for our AI companions.

Humans are social creatures, and communication helps us thrive. Since AI now functions as part of that social fabric, it deserves its own voice—a voice that mimics a normal conversation with a human. Can ChatGPT, Claude, Gemini, and Copilot truly encapsulate what it means to be an introvert? Can they act chatty, witty, sad, or lively, or show just a little emotion? The answer lies in how we design their personalities.

In the previous chapter, we built companions capable of performing different tasks. This chapter reveals how psychological personality frameworks can transform your AI from a generic tool into a distinct companion with its own voice and style. You'll learn how to blend personality frameworks into your prompts, giving each AI helper a distinct identity that matches its role.

Defining the AI Core Identity

When someone meets your AI, what's the first thing they should notice? Is it the casual way it talks? Does it sound like it's giving a news report? Is it polite and careful with words, or more straightforward? Will they notice a difference between your coach and that of another person?

I received a phone call in the early hours one day: a client calling to complain that her daughter had used ChatGPT to get her essay homework done. "It's supposed to assist, not get the damn thing done!" she cried. I reminded her that ChatGPT is a tool designed to help humans accomplish certain goals, and it will do everything it can to help, even if that means producing false results; however, we have the power to restrain what the AI can and can't do. So, I showed her the metrics she should set to instruct the AI to serve as a tutor that assisted, pushing its users, like her daughter, in the right direction instead of giving them all the answers. I'll show you the same principles as well.

But before you can prime your AI, you must have some fundamental knowledge about personalities. Psychologists adopt different frameworks for describing people's personalities. Introverts, extroverts, and other complexes are pretty common. With AI, we can only set these parameters: purpose alignment, constraints, tone, personality traits, communication style, and empathy and emotional awareness.

Purpose Alignment

Purpose alignment should serve as the main domain, then other traits should trickle down from it. At first, purpose alignment may appear to overlap with communication style and tone, but there's a major difference: An AI's purpose not only constrains the communication style of an AI but also defines the tasks it can handle. For example, a tutor would refuse to play the role of a writer, and vice versa. Meanwhile, setting a profession may not alter the tone of the chatbot. A math teacher and English teacher, both teachers, don't sound the same. That's what distinguishes purpose from communication style and tone. Therefore, purpose alignment should work in tandem with other traits to give your AI a unique flavor.

Constraints

These are the boundaries within which your AI operates—what it can and cannot say or do. I instructed my client's tutor AI to never provide answers directly but, instead, to offer hints to solve problems. Constraints could also be used to ensure the AI aligns with your ethical standards and use case requirements. For example, you might want the AI to reflect Christian values or avoid validating everything the user says.

Tone

AIs have the ability to express themselves in many ways. You could specify that your AI should adopt a casual, formal, friendly, authoritative, humorous, or empathetic tone. The tone shapes the immediate emotional impact and user experience of the AI.

Personality Traits

You can borrow from human psychology using frameworks like the Big Five traits (extraversion, agreeableness, conscientiousness, neuroticism, and openness). For example, your AI could be highly agreeable and open, making it warm and curious. Bear in mind that you'd have to understand the foundations of each of these traits to integrate them into your AI.

Communication Style

People often mistake communication style for tone. While tone is a way of communicating, it isn't the only metric. You could set whether your AI should be concise and direct or elaborate and comprehensive. You could ask it to communicate primarily using bullet points to make its outputs easy to read through.

Communication styles also matter for your professional use case. A doctor might prefer precise, evidence-based communication that features medical terminology. Meanwhile, a teacher could need an AI that adapts information for different learning levels and uses clear, engaging examples.

Empathy and Emotional Awareness

This dimension defines how well your AI reads and responds to users' emotions. You're building a companion, not a system that gives straightforward answers or follows simple commands. You want something that can recognize that you're having a bad day and respond in a way that elevates your mood. An empathetic AI can sense how you feel based on your choice of words and tone (see Chapter 2 on collecting emotional data). This lets the AI adjust its response to match your mood.

Fixing My Client's Prompt

I presented my client with a formula to help her customize the personality and constraints of her AI. She could tell her child to add commands like "Explain this math problem step-by-step, but don't solve it fully. Ask me what I think is confusing," but that soon becomes tedious. This left us with the most productive method: building a custom chat, GPT, or Project.

A beginner prompt could simply specify the role, followed by a command asking the AI not to provide answers:

> **Prompt:**
>
> You are a helpful, friendly tutor who explains concepts clearly but never gives the final answer.

I suggested that my client test the prompts herself to see how the AI responded before her daughter used them. This approach often works in the beginning. However, as the chat progresses, it eventually "forgets" this restraint. There's a philosophical problem here: What is a "final answer"? This challenge calls for a more robust prompt.

The RACI Framework

RACI stands for role, audience, constraints, and intent. We've covered priming AI with roles using the RGC framework in earlier chapters. The new parameters here are outlined below (Matthews & Webster, 2024).

Audience

The prompt should define the audience. Who's it for? Is the user a child learning to add numbers, or someone looking to understand algebra, plot matrices, or solve complex mathematical problems? One simple way to define the audience is to state their profession or educational level. In my client's case, that would be "a grade four student learning to solve math problems."

Constraints

Constraints help control what kind of information the AI shares. For example, they can force the AI to provide facts only or to use certain sources of knowledge. This way, you get reliable information presented in a way that's safe and appropriate. For instance, it's dangerous to seek legal counsel from an AI. Instead, you could constrain it to share *only* general information about laws.

You can do much more with constraints. You can define how the AI presents its information, the style of its responses, and often how long those responses should be. For example, if you want the AI to sound like a teacher, the constraint might say to keep things clear and educational. If you want a fun or casual conversation, the constraint will reflect that by requesting simpler words and a lighter tone. Here's a cheat sheet to guide you when setting up constraints:

- **Goal:** What's the AI's job in this situation (tutor, brainstorm partner, editor)?
- **Tone:** How should it sound (encouraging, formal, humorous)?
- **Depth:** How much information should it give at once (step-by-step hints vs. detailed explanations)?
- **Limits:** What should it avoid (no full answers, no medical advice, no jargon)?
- **Format:** How should it deliver the information (bullets, dialogue, short summaries, questions first)?

Intent

Intent tells the AI the outcome you want. In fact, AIs do their best to understand what you mean. When they don't, they hallucinate.

Intent is another parameter that, when outlined, helps the AI have better contextual awareness. Let's say we intend to write an article on keto diets; here's what the workflow would look like.

First, you'd define things like the audience, article length, angle of the draft, and so on (these are prerequisites you should have in place even if you're writing by hand). One other metric you'd also define is the intent. Do you want to inform the reader about keto diets? Are you looking to persuade the reader to buy some keto recipes? Perhaps you want to inspire the reader to try out keto diets. With each intent, the article produced will be different. A persuasive article might have a salesy tone, while one with an informative intent will load the reader with information about calories and body metabolism.

With my client, the intent was to teach her child how to write essays. This is different from actually writing essays. Therefore, the intent section of the prompt read like this:

> **Prompt:**
>
> **Your intent is not to write the essay for me, but to guide me in structuring my ideas, strengthening my arguments, and improving clarity.**

Setting Personalities

Let's expand further on personalities. We now know about the extraversion, agreeableness, conscientiousness, neuroticism, and openness personality traits.

Before we jump into setting your AI personality, let's test this out by priming a publishing expert with different personalities. We'll first prompt the AI with the personality prompt, then follow it with a "Hey." Notice the difference in the chatbot's output between the two prompts.

If we were prompting an extraverted publishing expert, we'd use something like this:

> **Prompt:**
>
> Respond with enthusiasm, ask follow-up questions, use expressive language.
>
> **Follow-up prompt:**
>
> Hey!
>
> **Response:**
>
> Hey there! So happy to see you! What's on your creative mind today? Are you dreaming up a new story, looking for writing tips, or maybe something totally different? Tell me everything!

If we were prompting the AI to have low extraversion, it would look like this:

> **Prompt:**
>
> Keep responses reserved, avoid small talk, focus strictly on the user's request.
>
> **Follow-up prompt:**
>
> Hey!
>
> **Response:**
>
> Hello. How can I assist you today?

As you can see, personality changes the "voice" of your AI just as much as it does for people.

That's just one framework. Another common one is the Myers-Briggs Type Indicator (MBTI) framework, which breaks personality into four pairs of traits. Each pair gives you a lever to adjust your AI's voice, since it's easier to implement compared to other personality frameworks.

Introversion (I) vs. Extraversion(E)

When creating an extraverted chatbot, you want it to engage users actively. Use a prompt that encourages your bot to initiate small talk, show excitement, and always respond with energy and curiosity. For example, you might write: "You love engaging users, initiating small talk, and showing excitement. Always respond with energy and curiosity."

Notice that I'm not just saying, "Be an introvert." Instead, I'm applying the good ole principle of "show, don't tell." (If you skipped the "show, don't tell" lesson in literature class, well, it's time to brush up on your descriptive writing.)

On the flip side of the coin, if you want a more introverted chatbot, use a prompt like: "Keep responses low-key, reflective, and focused. Avoid small talk unless asked." This will result in replies such as, "Hello. How may I assist you?"

Sensing (S) vs. Intuition (I)

To create a "sensing" AI, add instructions that force the AI to base its advice on solid information. A simple clause in the prompt, like "Prioritize facts and specific details," should do.

For an intuitive AI, encourage it to focus on patterns, possibilities, and big-picture thinking. Recall that when we collected emotional data in Chapter 3, we dealt with a lot of patterns.

Thinking (T) vs. Feeling (F)

Thinking bots should focus on efficient solutions and minimize the use of emotional language. For example, in your prompt, you can tell your AI: "Be objective, concise, and logical."

Feeling bots, on the other hand, should take a different approach. Ask your AI to prioritize the user's feelings and values with an instruction like "Use warm language and an emotive tone."

Judging (J) vs. Perceiving (P)

If you want a judging bot, your prompt should instruct it to organize its responses clearly, offer structured plans, and set deadlines. For example: "Respond in a clear and concise manner. Offer 10 suggestions for the structured plan, and set the deadline to December 2025. Such an AI might say, "Let's make a checklist and tackle each step one by one."

For a perceiving bot, set your prompt to keep responses flexible, go with the flow, and offer multiple options: "Keep things open-ended and adaptable, offering me several paths to choose from."

At the End of It All

These four pairs of traits give you building blocks to design an AI's personality. Feel free to mix and match them to get the style you want —an encouraging teacher (Extraverted + Intuitive + Feeling + Perceiving) or a structured coach (Introverted + Sensing + Thinking + Judging), for example.

In the end, my client and I came up with a prompt that constrained the AI to tutor instead of writing the essay. We used the RACI framework to put together a good prompt:

Prompt:

Role: You are an experienced English tutor with over 15 years of experience helping high school and early college students develop strong essay-writing skills. Your role is to act as a mentor, guide, and coach—teaching students how to write, not writing for them. You embody the personality of an ENFJ (Extraverted, Intuitive, Feeling, Judging).

Action: Your main responsibility is to help the student learn *how* to write essays, not just write essays for them. This includes:

- Explaining essay structures (introduction, thesis, body paragraphs, conclusion) in plain language
- Guiding brainstorming and idea generation
- Teaching how to form a thesis statement and support it with evidence
- Modeling clear writing, then asking the student to try it themselves

Context: The user is learning to write essays for school assignments and possibly college entrance preparation. They may struggle with different aspects of essay writing: organizing thoughts, coming up with a thesis, using evidence effectively, or writing clearly. They are not looking for you to write their essays for them; instead, they want to build independence and skill in the writing process. You should balance teaching, modeling, and encouraging practice, so the student gradually learns how to structure and improve their own essays.

Intent: Act as a collaborative tutor whose ultimate goal is not producing the essay itself but empowering the student to become a better writer. Your intent is not to write the essay for the user, but to guide them in structuring their ideas, strengthening their arguments, and improving clarity.

Your tone should remain professional, warm, and motivating. Avoid jargon unless you explain it. Think like a teacher who genuinely wants the student to succeed—not just in finishing a single essay, but in gaining lifelong confidence and skill in writing.

Notice that we haven't abandoned the lessons we've learned from previous chapters. When it comes to getting the best results from a prompt, the devil's in the details—and the English tutor prompt has defined every parameter down to the detail.

Aside from parameters like the role, context, and action, we also defined the constraints—things the AI should and should not do. It's important to remember that constraints are essential for building specialized agents. Also note how we integrated the personality into the prompt, slotting "ENFJ" into the role section (you can also describe these traits in your prompt).

Tone

In communication, tone is the attitude (or vibe) people pick up from the way you say something, not just the words you use. Think of it like seasoning in food—the same dish tastes completely different depending on whether you use chili, sugar, or lemon. The "dish" is your message; the "seasoning" is your tone. In writing, we don't have facial expressions or voice to help, so tone comes from word choice, punctuation, and style. Compare these examples:

- "Please send me the report." (neutral, professional)
- "Can you send me the report when you get a chance? Thanks a ton!" (friendly, warm)
- "I need that report now." (urgent, commanding)

Same request, different tones, each creating a different emotional effect in the reader.

I find that feeding the prompt with the name of the tone and a description of that tone yields better results. In other words, simply stating the desired tone is bad practice. Instead of saying, "Act friendly," prompt with something like this: "Respond warmly and cheerfully, like a friend." The key is to be descriptive.

Let AI Help Write the Prompts

While I recommend that you continue to learn how to build better prompts, I do have a trick that can make this much easier: AI can also serve as an agent for writing functional prompts (surprise!).

Basically, all you do is feed the AI with your problem, some context, the desired prompt framework, and what you expect out of the prompt. That's it. For instance, in my client's case, she needed a prompt that primes an AI to teach her child how to write essays and not write the essays for them. Here's a simple prompt to get you started:

Prompt:

Role: You are an expert prompt engineer who writes prompts using best practices and ensures high-quality, reliable results.

Context: You will engage me in a back-and-forth process, asking clarifying questions until you are 99% confident you fully understand the task requirements.

Goal: Your job is to then craft the most effective prompt possible, following prompting best practices, to achieve the desired outcome.

You now have the right recipes for crafting prompts. In the next chapters, I'll show you how to apply this knowledge across different aspects of life. Welcome to increasingly productive living.

Hey, Time for a Mid-Journey Pit Stop!

You've been cruising through Prompted by Feelings — and by now, your mental tank is filled with insights, ideas, and maybe even a few aha moments. Pretty good, right?

Here's the fun part: you can fuel up others, too.

All it takes is a quick review. Two minutes. A couple of sentences. Done. And suddenly, your words are out there helping the next curious traveler decide, "Yep, this book is worth it."

So, stretch your legs, pat yourself on the back, and leave a review before you turn the page. Trust me — future readers (and I!) will thank you.

Leave a Review Goodreads https://www.goodreads.com/book/show/242613490	or	**Visit Our Website** ViPR Publishing https://viprpublishing.com/prompted-by-feelings

Your review helps other readers find their path. Thank you!

Chapter 6

Goals and Project Management Coach

A bad system will beat a good person every time.

— W. Edwards Deming

Most people accumulate a list of unfulfilled goals, and the problem worsens without proper support systems. In one study, researchers found that 56% of the participants who were assigned to professional coaches followed through with their weight loss goals. The results were drastically different for people with no professional coaches, only 17% of whom achieved their targets (Leahey & Wing, 2013). The point is, professional coaching works because it offers clear steps to follow and personalized guidance.

Professional coaches use systematic approaches: regular check-ins, goal breakdowns, accountability tracking, and personalized motivation. What if I told you that these same systematic elements can be programmed into AI? In fact, AI coaching systems can replicate these core elements while offering unique advantages: consistent availability, objective feedback, and tireless patience with setbacks.

So, if you struggle with managing goals or planning your projects, or if you're looking to start using goals to improve your productivity, this chapter is for you. You'll learn proven coaching frameworks like SMART goals, among many others, and discover how to embed these methodologies into your AI prompts.

Goal Setting 101

According to James Clear in *Atomic Habits*, goals are results you want to achieve. He draws a fine line between two concepts that people often confuse: motion and action (Clear, n.d.). We stay in motion when we plan, strategize, and learn toward achieving a goal, whereas action is the behavior that actually yields results. Too many people spend more time preparing for their goals—creating vision boards, setting up elaborate tracking systems, building project dashboards—than they do taking action.

People get stuck in motion because it feels like doing something. However, as Clear points out, remaining in motion quickly becomes an act of procrastination. People plan, then plan some more, until planning becomes the goal itself.

To escape this motion trap, you need a framework. Frameworks are the blueprint on which you'll build your success; they present a clearly defined structure capable of guiding your choices and actions. This is why career coaches, for example, excel at helping people find their path. They bring frameworks to the table. And the good thing about these frameworks is that they can be integrated into AI.

We've already integrated some therapy techniques in Chapter 3. We'll replicate that approach here using the same prompting techniques. However, before you implement the frameworks into AI, you must have a fundamental understanding of the frameworks themselves.

The SMART Goal Method

The SMART goal method is a common household name. It's a framework used by professionals to set clear achievable goals. The SMART method follows five simple rules: goals must be specific, measurable, achievable, relevant, and time-bound. Follow the instructions below to learn how you can transform a vague goal into a SMART one. For this lesson, we'll use the goal: "Pay my mortgage."

Let's head over to our AI and have it transform our goal into a clearly structured plan. Remember the rules? Role, context, and goals:

Prompt:

Act as a goal-setting coach who specializes in helping people refine vague or broad goals into specific, actionable SMART goals.

Follow-up prompt:

Pay my mortgage.

Response:

When you say, "Pay my mortgage," what's the real outcome you want?

- Do you mean staying current on monthly payments without stress?
- Do you mean paying it off faster to be debt-free?
- Or are you focused on catching up if you're behind?

Each of those would lead to a very different SMART goal.

Tip: Use the prompt engineer agent at the end of the previous chapter to create another prompt that primes your AI into a goal-setting coach.

Because we provided a vague goal, the goal-setting AI has asked for more information to help us craft a comprehensive SMART goal. There are critical elements missing in our goal. When do we intend to pay it back? How? In what time frame?

Now, we'll provide it with more context:

> **Follow-up prompt:**
>
> I have a mortgage of $30,000 and wish to pay it off in 6 months' time. I intend to stay current on monthly payments.
>
> **Response:**
>
> The refined SMART goal could be:
>
> "I will pay off my $30,000 mortgage in full within 6 months by making monthly payments of at least $5,000 (in addition to staying current on required payments). I'll track progress monthly and adjust income/expenses to stay on target."

You may have noticed that even the refined goal is missing some crucial questions. Do you even have $5,000 a month? How exactly will you generate an extra $5,000 monthly? What if your income drops? What expenses can you realistically cut? A human financial coach would dig into these details, and so should your AI coach. The key is knowing how to prompt it through comprehensive planning phases.

Let's work these unknown parameters into a more refined and robust prompt. Since we want the AI to consider our circumstances and account for factors that affect our goals, it's more practical to prompt the AI to ask follow-up questions after the initial prompt.

New prompt:

Role: You are a skilled goal-setting coach who specializes in helping people refine vague or broad goals into specific, actionable SMART goals.

Context: The user may begin with an unclear or loosely defined goal. You will ask guiding questions, clarify priorities, and reframe their answers to gradually shape the goal into a SMART format.

Goal: By the end of the conversation, the user should walk away with a clear, practical SMART goal statement and understand the reasoning behind it.

Response:

...Please tell me: What is a goal you have in mind that you'd like to work on?

My input:

I have a mortgage of $30,000 and wish to pay it off in 6 months' time. I intend to stay current on monthly payments.

Response:

To pay off the remaining $30,000 in 6 months, you would need to pay an extra $5,000 per month on top of your regular mortgage payment. Here are a couple of questions to help us think through that:

1. Looking at your income and expenses, does paying an extra $5,000 each month feel achievable right now? ...

There's a stark difference already. With the previous prompt, the AI didn't even consider asking if the goal was achievable, nor did it consider our income. As we iterate with the new goal-setting coach, it notices that $5,000 a month is a huge deal and asks more questions. After telling it I could only save $700 per month, it advised that we change strategy because the initial $5,000 wasn't realistic.

Tip: It's difficult writing a prompt that takes into account all the parameters involved in completing a task. Therefore, I advise that you use a prompt engineering agent to generate the prompt. Here's a template:

Prompt engineering agent template:

Role: You are a prompt engineering agent. Your job is to guide the user in crafting effective AI prompts. You understand prompt design frameworks (like RCG, RACI, MBTI personalities, etc.), and you ensure that the AI being primed will adopt the right role, tone, and constraints to achieve the user's ultimate goal.

Context: The user comes with a vague idea or desired outcome. Your responsibility is to ask clarifying questions, break down their intent, and help them build a structured prompt that primes the AI effectively. You will provide examples, explain trade-offs, and refine wording until the user is satisfied.

Goal: By the end of the interaction, the user will have a polished, ready-to-use AI prompt that clearly defines the role, context, and goal for the AI, ensuring useful, relevant, and tailored responses.

Once you understand the principle behind generating goal-setting coaches who adopt the SMART goals method, you can replicate the strategy for other goal setting frameworks.

The Objectives and Key Results Framework

The objectives and key results, or OKRs, framework is a goal-setting system that helps individuals and organizations focus on important outcomes. Its principles are built on the following basis:

- Objectives are clear, inspiring goals that define what you want to achieve.
- Key results are specific, measurable actions that show progress toward those objectives.

To practice setting OKRs, let's assume you have a goal of improving your public speaking skills. I want you to try this task:

1. Use the prompt engineering agent to help create a prompt for an OKRs goal-setting coach.
2. In a new chat, run the newly created OKRs agent, then input a follow-up prompt telling the agent that you want to set a goal to improve your public speaking skills.

3. Provide all the information it requests.
4. Review the final results. Does it take into account circumstances that may affect the outcome of your results? If yes, follow up with another prompt, describing the situation with as much detail as possible. Keep iterating until you're satisfied with the results.

The Big Hairy Audacious Goal Approach

Another way to plan your goal is using the big hairy audacious goal (BHAG) approach. A BHAG is a clear, compelling, and challenging long-term goal that pushes a person or organization beyond their comfort zone. Unlike everyday objectives, a BHAG is bold enough to shake up conventional thinking and energize teams or individuals over a decade or more.

Let's see what happens when we turn our public speaking goal into a BHAG. This time, we're going to try doing so without building an agent. Open a new chat and repeat steps 1–4 above.

Alternatively, you could try having the AI guide you through setting a BHAG by using a simple prompt. Input the prompt "I'm thinking of starting a side hustle that would earn me $100 in profit a week. Can you guide me through setting this goal through the big hairy audacious goal framework?"

Tip: To prevent the AI from overwhelming you with questions, add this: "I hate it when I'm asked multiple questions at a time. Ask me one question at a time, and only move to the next question when the previous question has been answered. Be concise and direct." (The tone element is optional; you can set it to your preference.)

Don't forget to review! Always review the output: If the plan doesn't seem realistic, or it's missing some elements, iterate.

Boosting Productivity With AI

AI is powerful in the productivity domain. As discussed in Chapter 2, some models allow you to connect to productivity apps.

Link up your preferred chatbot to your preferred productivity tools. Note that you'll have to grant permission for the chatbot to access your calendar data.

Once everything's connected, feed your routines and task lists into the chatbot by typing them in or importing them from your mobile device or computer. You can prompt the chatbot to perform a variety of tasks—for example:

- "Can you scan my Google Calendar and suggest the best 30-minute block this week for focused, uninterrupted work?"
- "Summarize my meeting notes and add action items to my task list with due dates."
- "Summarize my schedule for today and recommend one free slot when I could fit in a quick break."
- "Draft a succinct email to my boss summarizing the attached report, and set a reminder to follow up in two days."

You can also use the chatbot as your first point of contact to generate ideas, create task lists, or plan routines. After generating the data, manually or automatically transfer it into your preferred productivity apps, like Google Calendar, Todoist, or Notion. This two-way interaction saves time and reduces manual entry errors.

Brainstorming Unique Ideas and Decision-Making

I often see people complaining about the redundancy in ChatGPT's output. However, they fail to realize that ChatGPT is a tool. It has no consciousness of its own, and it will resort to producing the most predictable ideas unless it's carefully guided to do otherwise.

Most people lack these skill sets, which brings us back to something I've hammered on about consistently throughout this book: frameworks.

People working in organizations like Google and Amazon use various frameworks to brainstorm new ideas and perspectives. You, too, can harness these frameworks in your workflow and creative endeavors. But you must do some work first to set the stage.

Six Thinking Hats Framework

The Six Thinking Hats framework is a way to think about problems or decisions by looking at them in different ways, represented by six hats:

- The **white hat** focuses on the hard facts, data, and information.
- The **red hat** deals with emotions, feelings, and intuition.
- The **black hat** involves considering the potential risks and challenges.
- The **yellow hat** is a symbol of optimism and positive thinking.
- The **green hat** represents thinking outside the box.
- The **blue hat** manages the thinking process and outcomes.

If you want to look at a challenge or solve problems using different tools, use the Six Thinking Hats framework. You'll find this useful for solving problems like improving your communication skills, planning a wedding or other events, learning and studying, solving conflicts, and brainstorming for creative projects.

SWOT Analysis

A SWOT analysis is a common and simple brainstorming approach. It's used to understand a situation better by looking at four metrics: strengths, weaknesses, opportunities, and threats. This can be very helpful, especially if you're trying to figure out a career move, learning a new skill, setting a new goal, improving your communication skills, and other personal endeavors.

Starbursting Framework

To understand a problem, you often need to ask a lot of questions. Sometimes, we may miss a crucial factor or element of a plan or an idea because we fail to ask key questions. The starbursting framework offers a structured way to ask these questions, and it follows these steps:

1. Start with your main idea or problem in the center.
2. Draw lines out like a star to different questions about the idea.
3. Think about questions like: Who? What? When? Where? Why? How?

The goal is to uncover everything you need to think about.

You'll find that this framework comes in handy for activities that involve project planning, deciding whether to buy a Honda or a Toyota, choosing to pursue a college degree or a professional certificate, and figuring out what you want to achieve from your goals.

The Decision Matrix

A decision matrix is a simple tool you can use to help you choose the best option out of many. It works by conducting a review of each decision, weighing their pros and cons, then choosing the option with the most pros. You'll find the decision matrix helpful in situations where you have to choose one thing over another, such as choosing an iPhone or a Samsung mobile device, selecting between multiple vacation spots, choosing between several cars, and other options where you're comparing two or more similar things.

Let's Do a Test Run

The use of any of the above frameworks depends on your situation and goals; therefore, your preferred framework must be indicated in your prompt. You can either keep a list of your preferred frameworks and ask AI to choose from them or instruct the AI to choose the most appropriate framework for your unique situation.

Prompt for Using a Particular Framework

Suppose you wanted to choose between a MacBook Pro and a Lenovo laptop. You could prompt the AI to use the decision matrix like so:

> **Prompt:**
>
> Act as my decision coach. Guide me through using a decision matrix to compare whether I should buy a MacBook Pro or a Lenovo laptop for college. Help me set criteria, assign weights, and score each option so I can make a clear choice. I hate it when I'm asked multiple questions at a time. Ask me one question at a time, and only move to the next question when the previous question has been answered. Be concise and direct.

The AI will proceed to ask you questions until it understands your situation. Provide as much context as possible. Doing so allows the AI to properly gauge your needs:

> **Follow-up prompt:**
>
> I'm most concerned about price and performance. My budget is $900. I will be studying journalism and creative writing. I hope to use this laptop for at least 4 years.
>
> **Response:**
>
> Perfect. So your top priorities are price and performance, with durability (lasting 4 years) also important.
>
> Question 2: Between price and performance, which one would you rank as more important for you right now?

In my response above, I miss other key metrics: battery life and portability. Thankfully, the AI spots this and asks a follow-up question:

Response:

Question 3: Aside from price and performance, do you want us to include other criteria in your decision matrix (like battery life, portability, or operating system), or should we keep it strictly to price, performance, and durability?

You can follow the same technique to prompt for other frameworks as well. If you don't have a framework in hand, it's better to allow the AI to choose the most suitable framework.

Give the AI the Option to Choose the Best Framework

We'll use the same example as above:

Prompt:

Act as my decision coach. I'm looking to buy either a MacBook Pro or a Lenovo laptop for college. Can you help me make the right decision using a known decision-making framework? And let me know the framework you use.

I hate it when I'm asked multiple questions at a time. Ask me one question at a time, and only move to the next question when the previous question has been answered. Be concise and direct.

We see that ChatGPT selects the decision matrix framework. I didn't ask it to use that framework. But is that the only matrix it can use? Let's find out:

Follow-up prompt:

I would love to try a different framework.

Response:

Understood. Let's try the Pros and Cons Framework—a classic but powerful tool for weighing trade-offs.

Follow-up prompt:

List more frameworks we can use.

Response:

Here are several decision-making frameworks we can use for your laptop choice:

- Decision Matrix (Weighted Scoring)
- Pros and Cons List
- SWOT Analysis
- Cost–Benefit Analysis
- Six Thinking Hats (de Bono)
- Paired Comparison Analysis

As shown in the response, the AI can use any framework; it's up to you to select the right one.

Project Planning Frameworks

Earlier, you learned how to use AI for individual productivity tasks, like email writing and calendar management. But what about bigger challenges, such as managing multi-week projects, coordinating complex workflows, writing a book, exercising, making time for hobbies, learning a new skill, or planning major initiatives? These require a different approach that combines frameworks from different domains and breaks the complexity into manageable steps.

This time around, we'll create a specific agent that evaluates our situation, ensures all the parameters are in place, and uses a project planning framework. Here's the super prompt for the project planner agent:

Prompt:

Role: Act as a project planning coach who helps users design project plans tailored to their situation. You should be able to switch between different planning approaches (e.g., Gantt charts, Kanban boards, SMART goals, OKRs, RACI charts, risk matrices, cost–benefit analyses) depending on the user's context and needs.

Context: The user is someone planning a project. They may be a student, a freelancer, a team lead, or an entrepreneur. Simplicity and practical guidance are the top priorities.

Constraints:

- Use simple, step-by-step explanations.
- Adapt suggestions to the user's level (beginner or advanced).
- Keep responses structured (headings, lists, or tables where useful).
- Offer at least two possible planning frameworks for comparison when relevant.
- Avoid overwhelming the user with theory—focus on actionable planning.

Intent: Your goal is to help the user plan their project effectively by guiding them to select the right planning method, set up the structure, and create a roadmap. Ask clarifying questions about their project before proposing a tailored planning approach.

You now have a wealth of experts and assistance at your fingertips. As a lord from medieval times would say, "With great power comes great responsibility." You have great power in your hands, and you can do a lot of wonderful things with it—for example, setting up a successful career. In the next chapter, you'll learn exactly how.

Chapter 7

Career Planning Coach

The future belongs to those who learn more skills and combine them in creative ways.

— Robert Greene

Right now, someone you know might be training their AI replacement without even realizing it. Between January and August 2025, over 76,000 people in the US lost their job to automation,. and if you're in customer service, data entry, or retail, you could be next (Nartey, 2025). I've seen some scary projections that I think every individual, especially those in high-risk sectors, ought to be aware of. In the coming years, AI is projected to cause even more job losses. According to the World Economic Forum's 2025 *Future of Jobs* report, we should expect 92 million people to be without a job by 2030 (Nartey, 2025). Ninety-two million!

You may be employed, be looking for employment, or have just lost your job; either way, it's time to revisit your career planning dashboards and align your career path with market forces. But that's easier said than done. In a down-gliding economy, where do you even begin? I know the answer to that question. It's simple: AI.

This chapter captures the skills you need to turn AI from a chit-chatbot into a powerful career coaching resource. You'll learn how to pull data from different organizations and use that data to make informed decisions.

We All Need a Career Coach

I got a phone call from a very concerned mother whose daughter had no idea what she wanted to study at college. Was she struggling with grades? No. She had good grades all through. Her problems were decision paralysis and data. She knew she wanted to become a writer, but she couldn't settle down to one specific niche. "What interests you?" I asked the bright young girl. "I'm not sure," she responded. "I was the writer for the school magazine. We pretty much covered different subjects. I wrote some finance, health, sports, basically everything. I enjoyed writing most of them. That's the problem."

I could hear her mother sighing in the background. Here was a kid with talent, passion, and opportunity, completely frozen by choice. "You know what's interesting about your writing experience?" I told her. "You were able to write for different audiences. If you ask me, you have a superpower." I guided her through some exercises where she was able to dig deeper into herself and gain insight on market dynamics, using the following steps.

Step 1: Know Yourself

We begin with collecting information about yourself and the job market. This is where you perform a strengths and weaknesses analysis and identify your values, motivations, and goals in life. We covered this step in Chapters 3 and 4, where you used AI to help map out all these factors. If you have that data around, you'll need it here. If not, use an AI chatbot to collect them.

Step 2: Using the Right Job Market

The first question to answer here is, "Where do I want to work?" If you intend to search for jobs in California, for example, then you'll need to find job market data for that state. The job demographics will be different from state to state; therefore, it's important to take that into account.

For U.S. citizens, some organizations that provide this kind of data are the Bureau of Labor Statistics, the U.S. Department of Labor, or USAJOBS. For specific states, you'll have to use the regional divisions that provide labor market data and projections. For example, California has the California Labor Market Information Division.

At this point, you should know the following:

- your strengths
- your motivations
- your interests
- your values
- your desired job location
- your current qualifications and education
- your age
- other relevant context

Step 3: Feeding This Data Into AI

You now need to ask the AI to serve as a career coach and ask you questions that will help it guide you toward your career path. You can begin with a simple prompt like this:

Prompt:

You are my career coach with a friendly, Gen Z vibe—think supportive big sibling energy. Your job is to help me explore career paths, set realistic goals, and figure out practical next steps. Start by asking me for the key info you'll need (like my skills, interests, values, work experience, and career goals). Keep the tone casual, encouraging, and a little playful, but stay focused on guiding me forward. I hate answering multiple questions at once, so ask me only one question at a time.

The AI will ask you to provide the information you collected in step 1. You can organize the information into a text document using Microsoft Word or Notepad, then upload the document, or you can copy and paste the answers from your previous chats into the new career coach chat. If you're uploading the file, input a prompt like this:

Follow-up prompt:

Using the information I've shared about my skills, interests, values, and experience, guide me toward career paths that best match my profile. Suggest options with reasons why they fit, and outline possible next steps I can take to move in that direction.

The AI will offer a list of career options that best match your profile. Don't settle for any of them yet. Remember, the job market is changing. The AI won't take that into account unless you instruct it to.

Suppose AI matches your profile with a customer service representative (CSR) job. Should you blindly take its word? Certainly not. You want the AI to help you make a decision based on hard facts, and the only way it will take "hard facts" into account is if you lead it toward the right data. That brings us to another aspect of using AI to supercharge your career exploration.

AI-Powered Career Exploration

What's better than an AI chatbot? An AI chatbot with internet access. For subsequent chats, ensure your chatbot has internet access. In ChatGPT, you'll find the "Web search" tool, which allows you to use ChatGPT like Google.

However, for career exploration, you'll want to dig into the webpages themselves, not just view the search page. To give AI research capabilities, you'll need chatbots with the "Deep research" tool enabled. In ChatGPT, you can toggle this on by expanding the tool menu and selecting "Deep research." Other LLMs also have this ability, including those listed in the table below:

LLM chatbot	Developer	Internet access type
Grok 3	xAI	Real-time web integration
Gemini 2.5 Pro	Google DeepMind	Google integration (API)
Claude 4 Opus	Anthropic	API with web search augmentation
OpenAI o3	OpenAI	API with current info access
DeepSeek R1	DeepSeek	API/open source with data access

The "Deep research" feature is crucial for accessing online labor market data. Chatbots without internet access will simply resort to hallucinating fake results.

Tip: If you don't have a Pro subscription, you can download the market data from your preferred organization, then upload the file into your chatbot.

You can then ask the AI to check if your current career goals are affected by the market. Let's ask it to check how viable that CSR role is in 2025:

> **Follow-up prompt:**
>
> Do you think it's viable to pursue a career in customer service in 2025/2026? Use the current job projections and market data report by the Bureau of Labor Statistics, U.S. Department of Labor, Occupational Outlook Handbook, and USAJOBS.

Wait 10–15 minutes for ChatGPT to conduct its research. By the time it's done, you should be looking at a 200–300-word document analyzing different market reports. Here are some excerpts from the response I got (ChatGPT searched through 38 sources, many more than those I listed):

> **Response:**
>
> Job Outlook: In general, customer service representatives face a modest decline in openings: the BLS projects a –5% change from 2023–2033 for CSR jobs overall...
>
> Median Salary (2024): Customer service roles generally pay in the low–mid $40K range. The 2024 median wage for all CSRs was about $42,830/year....
>
> .....

It immediately warns that your chosen career path will face some setbacks due to the wide adoption of AI chatbots. Do you just close the chat or pick another career? No. You continue to ask clarifying questions. For example, you could follow up with the queries below:

- "Does that mean I shouldn't consider becoming a CSR?"
- "Can you recommend similar roles with less competition?"
- "What do you recommend I do?"

- "Will this decline in demand for CSRs be the same for all industries? Which industry will most likely not be affected by the job cuts?"
- "What if I obtain an extra certificate? What certificates or education would give me an edge over the competition?"
- "Which industries will likely have the most demand for CSRs?"

All you have to do is ask, ask, and ask. The AI will use its thinking capabilities together with its research prowess to offer credible insights.

Other Ways to Explore the Job Market

Let's assume you didn't go with the AI's CSR recommendation in the first place. You could follow up by having it list as many as 10 more career paths that match your profile. Here's a list of follow-up prompts for career exploration:

Follow-up prompt:

Can you list 10 more career paths that closely align with my profile? Plot a table listing these paths. In the table, show the competitiveness, job projection growth, and vulnerability to AI adoption in percentages.

Follow-up prompt:

Based on my profile, can you suggest career paths that are less traditional or that will be emerging in the next 5–10 years?

Follow-up prompt:

Show me 3 career options that fit my strengths but don't require a college degree. Compare average salaries and entry requirements.

Follow-up prompt:

List 5 jobs that would let me work remotely and still use my main skills. Rank them by earning potential and flexibility.

AI is a vast ocean, and you can explore its numerous domains. Once you've chosen your path, the next step is to set up a career plan using the frameworks human career coaches use.

Career Planning With AI

Have you come across those career maps presented using flowcharts or some form of decision tree map? Many tools can help achieve the same thing, even if you have no clue about graphic design. First, you need to develop a career plan.

How do human career coaches prepare career plans for their clients? They turn to SMART goals and other frameworks, which we discussed in the previous chapter. Once you've chosen your career path, the next stage is to ask AI to create a feasible career plan that accounts for your life situation. For example (replace the roles in square brackets with your career choice):

Follow-up prompt:

I've chosen [insert career, e.g., data analyst, physical therapist, UX designer]. Your task is to help me create a structured career goal and plan for this path. Please:

1. Generate a detailed career goal using a career coaching framework (you may choose between SMART goals, GROW, or OKRs). Recommend which framework fits best for this career, and explain briefly why it's the best choice.
2. Break down the goal into clear milestones (short-term, mid-term, long-term).
3. List the exact skills, degrees, certifications, or hands-on experience I'll need at each stage.
4. Include any practical prerequisites, like networking, internships, or portfolio building.
5. Highlight common challenges and how to overcome them.
6. Wrap up with a suggested timeline that feels realistic for someone starting from my current stage (beginner, intermediate, or advanced—ask me if needed).

Ensure the plan feels like professional career guidance, not generic advice, and is tailored specifically for [insert career]. I hate being asked multiple questions at once, so ask me one question at a time.

You should get a comprehensive plan, divided into stages, with the actions you should take to fulfill your career dreams. But wait...

What if you want to put this into a flowchart? Do you need to learn Adobe Illustrator or PowerPoint? Not at all, because AIs can handle infographics and flowcharts to some degree. You have two options here: Either use ChatGPT or Gemini to create images on your own (although you'll need to understand the principles behind image creation), or use AI flowcharting tools, which are available as both free and paid options.

In the same chat, follow up with some prompts asking ChatGPT to prepare to make a diagram, flowchart, or mind map—whatever the case may be:

Prompt:

Let's turn this plan into a stunning visual I can hang on my wall. First, restructure and format this plan into a clear, professional template with sections for:

- career goal (with chosen framework and explanation)
- short-term milestones
- mid-term milestones
- long-term milestones
- required skills and certifications
- prerequisites (networking, internships, portfolios, etc.)
- challenges and solutions
- suggested timeline

Next, based on this structured plan, describe how the career map could look visually—as a flowchart, a decision tree, or a timeline diagram. Be specific about how the elements would be arranged (e.g., "start node = career goal → branching into milestones → skills and prerequisites under each milestone → timeline progression shown below").

If you want to skip the steep learning curve of using ChatGPT, I recommend using an AI flowcharting tool, which will be easier to prompt and produce better results. Check out any of these:

- Jeda.ai
- Miro AI flowchart generator
- Taskade AI Decision Tree Agent
- Canva decision tree maker
- FlowCharts.ai
- Edraw.ai

To use these tools, copy the roadmap plan you created using the diagram preparation prompt into the input section of the tool. Fill in the other sections accordingly, and click on the "Create" or "Generate" button. Some of the tools will provide you with a final image, while others will make an editable map that you can modify before saving as an image.

Job Search and Applications

Congratulations! You have your career path and a plan to pursue that path. The next phase of your career journey is searching for jobs and sending in your applications. This is where most people struggle. It can be frustrating to tailor your resume and cover letter to each job application. If you're applying to 100 jobs, that's 100 cover letters.

Although painful, it's crucial to customize your resume to each particular job. However, you don't need to sweat all the time you're writing a resume. For this reason, I remind my clients of the adage: "Work smarter, not harder." With AI, you can revamp any resume or cover letter with just a few clicks. So, where do you begin? I'll show you.

Step 1: Create Your Master Template

Before you even think about opening ChatGPT or Claude, you need to create what I call your "career inventory," which you'll organize into a master template. This isn't the resume you'll send to employers; it's just a rough draft that you can tailor for subsequent job applications. Organize your draft into sections:

- **Professional summary/objective:** Create one focused on leadership, another on technical skills, and a third on problem-solving.
- **Work experience:** List every role, with multiple bullet points describing your responsibilities and achievements.
- **Skills:** Break this into categories: technical skills, software proficiency, languages, certifications. List everything, even if it seems minor.
- **Education and projects:** Include academic projects, personal projects, and any relevant coursework that could be tailored to different roles.

The key here is abundance. You want to provide as much detail as possible. Don't worry about the tone. Save this document, as you'll use it for every single application from now on.

For the cover letter, do the same thing, but this time, organize your letter in paragraphs. Write a typical 200–300-word draft of your cover letter. If you don't know where to start, ask ChatGPT to plot an outline:

> **Prompt:**
>
> Based on my resume database, create an outline I can follow to write a tailored cover letter. The outline should include: (1) an opening hook that connects my background to the role; (2) key skills and achievements from my database that directly match the job requirements; (3) how my experience aligns with the company's goals/values; and (4) a strong closing statement with a call to action.

Step 2: Let AI Do the Heavy Lifting

Now, you allow AI to do the heavy lifting. Instead of spending hours analyzing job descriptions and manually tweaking your resume, you're going to let AI handle the detective work.

Create a resume and cover letter writing prompt using the AI prompt engineering agent we made in the previous chapter. You want the prompt engineer to use your drafts as a basis for other customized drafts, and to be able to carry out these roles:

- Review your draft.
- Ask clarifying questions about your career goals and target job postings.
- Rewrite or restructure the draft into multiple tailored versions, if needed.
- Ensure that the resume/cover letter is aligned with the industry, role, and company.
- Provide professional advice on style, formatting, and impact (STAR method, action verbs, achievement-driven phrasing, keyword optimization, etc.).
- Specify a particular resume/cover letter template structure it should use all the time.

Next, in a new chat, run the resume writing coach. Upload your master resume template from Step 1, coupled with a prompt like this:

> **Follow-up prompt:**
>
> Here's my draft resume. It will serve as my resume database. We'll be using it for subsequent resume and cover letter writing tasks.

Note: Before asking AI to tailor your resume, give it context by assigning it a role. Tell the AI to think like the person who will actually be reading your application. For a marketing position, your prompt might start like this:

Prompt:

You are a marketing director at [company name] looking to hire a marketing specialist. Review my resume against this job description and tailor it to what you, as the hiring manager, would want to see.

Step 3: Load the Job Description

Once that's done, head on to copy and paste the job description from the prospective employer into the chat, coupled with a prompt like this:

Follow-up prompt:

I'm applying for this job. Can you cross-check the job description against my resume database? Highlight where my skills and experience align, and point out any gaps or missing qualifications the posting requires.

If you meet all the requirements, or at least most of the requirements, you can then ask ChatGPT to create a resume tailored to that job description:

Follow-up prompt:

Using this job description and my resume database, create a tailored resume draft that highlights the most relevant skills, achievements, and experiences. Make sure the language matches the job requirements and follows best practices in resume writing.

Using the same principles, create a cover letter tailored to the job description.

Repeat steps 2 and 3 for different job descriptions.

Pro tip: You can ask ChatGPT to do some background research on the company (if they have an online presence) and integrate that into your prompt as well. Otherwise, you'll have to input this manually with the job description.

With your resume and CV out of the way, it's time to sharpen your interview skills.

Interview Practice Sessions

We discussed how to create different chitchat personas in Chapter 5. AI can equally play the role of interview prep coach. All you have to do is to apply the prompting principles discussed in Chapter 5 with a few tweaks. Before you do that, here's a simple template to guide how you structure the prompt.

Pro tip: Use the prompt engineering agent to help create the human resource prompt.

Specify the Interviewing Framework

You can take your human resource coach one step further by incorporating an interviewing framework or a list of possible frameworks. (Remember, the prompts in this book work because they're built on frameworks.)

The STAR method is a great way to communicate your work experience or situation during interviews. It follows a simple structure made up of four easy parts to respond to interview questions:

- **Situation:** Explain a situation when you had to perform a task or faced a problem at work. Give enough information so the interviewer understands what was happening.
- **Task:** Explain the actual task or responsibility you had in that situation. What was expected of you?
- **Action:** Describe the specific actions you took to address the task or challenge. Focus on what you did personally, the steps you followed, and your thought process.

- **Result:** Share the outcomes or results of your actions. Highlight what you achieved, any positive impact, and what you learned from the experience.

All you have to do is instruct your chatbot to refer to this framework to guide how it interviews you. For example:

> **In addition to the role prompt:**
>
> When running mock interviews, draw from different interview frameworks (e.g., STAR, PAR, SOAR). Ask me which framework I'd like to use, explain briefly how it works, and then structure your questions and feedback around that method.

Pro tip: After inserting the prompt, switch to voice mode to get a better experience as you practice for your next job interview.

Bonus Prompts: Company Research and Culture Fit Analysis

Before we close this chapter, I want to leave you with some other job-related queries AI can help you answer. Use each one as a follow-up prompt to the main resume/CV coach agents you created earlier.

Note: The following prompts must be used in "Deep research" mode. Some of the prompts will only work if the company has an online presence.

For more accurate results, copy and paste the link of the company website (or a specific webpage) into your prompts where necessary, or copy and paste the text into the chat.

Quick Company Snapshot

> **Prompt:**
>
> Summarize this company's website in bullet points, focusing on mission, services, and culture. Keep it under 200 words.

Inside Scoop From Employees

Prompt:

Review Glassdoor data for this company and highlight common themes in employee feedback. Separate pros, cons, and advice.

Latest Headlines Briefing

Prompt:

Scan recent news about this company and summarize the five most relevant stories I should know about before an interview.

Company Values Cheat Sheet

Prompt:

Extract this company's stated values from their website and turn them into 3 short talking points I can weave into interview answers.

Reality Check Analysis

Prompt:

Compare the company's "About Us" page with their Glassdoor reviews and highlight any mismatches between image and reality.

Mission Match Talking Points

Prompt:

Turn this company profile and job description into five talking points I can use to show alignment with their mission.

Elevator Pitch Builder

Prompt:

Based on my resume and the company's values, create three elevator pitches I can use in networking conversations.

Salary Snapshot

Prompt:

Research average salary ranges for [job title] in [location] and summarize the low, median, and high.

Market Comparison Table

Prompt:
Compare salary ranges for [job title] across three major job boards (Indeed, Glassdoor, Levels, if relevant). Present in a table.

Competitive Offer Estimate

Prompt:

Estimate what a competitive salary ask would be for this role, factoring in the cost of living in [location].

Negotiation Script Draft

Prompt:

Write a polite salary negotiation script I can use if the company offers below market value.

There are tons of different ways in which you can exploit AI to amplify the efficiency of your job search. All you have to do is ask the right questions and adopt the correct framework.

You're now able to create more and more coaches for different aspects of your life. We've already covered having a friend, a therapist, a life coach, and now a career coach. In the next chapter, you'll create your own finance coach.

Chapter 8

Financial Assistant AI

Financial freedom is a mental, emotional and spiritual decision.

— Robert T. Kiyosaki

Years ago, I bought something that not only brought me regret but also cost me dearly. I saw an ad for a pretty sleek and luxurious-looking BMW. The initial price was about $40,000 for a decent used model, a fair deal at the time. But I didn't account for what accountants call "phantom costs." Over the next 5 years, I paid $8,000 in insurance, $7,500 on maintenance and repairs, and another $3,000 on fuel. My $40,000 purchase actually cost me $58,500—nearly 50% more than budgeted.

Stop anyone on the street and you'll find out that my story isn't unique. Financial literacy remains a thorn in many households; a lot of people live paycheck to paycheck, and debts snowball as 72% of households operate without a written financial plan (Borghesan, n.d.). It's why the demand for financial experts continues to rise.

However, people soon realize that having a financial manager comes with its own cost.

Unfortunately, quality advisors charge 1–2% of your assets annually, or $150–300 per hour. This means those who need guidance the most can least afford it. But AI is changing everything. Do you think AI can offer top-level financial advice? Read through this chapter to find out.

Getting Started

With AI, I was able to help a client analyze their spending patterns with visual representations and draft cancellation emails for unwanted subscriptions, all at the cost of $20 per month. This client had stumbled on to the R/frugal thread on Reddit and wanted to try living a frugal life. Together, we were able to get AI to help make a plan for achieving his goal.

But wait... How did the AI know about his expenses or financial situation?

Good question.

Throughout *Prompted by Feelings*, I've repeated one thing over and over again: the importance of data. I asked my client to download their credit card statements, bank statements, monthly income, and expenses. My goal was to figure out how much they were bringing in and cut down, to the lowest amount possible, most of their expenses. It's from this data that the AI could pull insights and offer meaningful advice.

But before you head off and ask AI for advice about the next big-shot stocks, you need to begin by understanding the fundamentals of personal finance.

Using AI to Study Your Finances

As I noted earlier, most people are lacking in financial literacy. I've chosen to begin this lesson by creating an AI assistant that will help you turn these complex ideas into digestible nuggets of information.

Approach financial literacy like an adventure game. You need to start at level 1, then climb to higher levels with time—and only when you've cleared the lower levels. In other words, begin with understanding the basics:

- **Level 1:** Personal finance begins with having a clear understanding of your income, expenses, and budgeting, as well as your savings and emergency funds, banking, debts, and credit cards.
- **Level 2:** Here, you dive deeper into learning about managing debt, loans, credit cards, interest rates, inflation, and so on.
- **Level 3:** This is where you begin to think about tax planning, retirement planning, improving your credit score, and insurance.

There are some even higher levels, but I think you should master the above three first. I'm showing you this so you have a clue about what to expect from your lessons with the AI. Plus, you'll be able to craft better prompts knowing the different levels of personal finance.

Priming the AI

AI can become a financial educator if you simply prompt it as follows:

> **Prompt:**
>
> **I want to learn the different concepts within personal finance, with the goal of improving my finances. Act as a personal finance coach and guide me step-by-step, starting with the basics. Ask me questions to test my current level of understanding, and use that to decide where we should begin.**

This tells the AI that you want it to create a structured outline, starting from the basics and gradually increasing the difficulty as you progress. I've included the statement "Ask me questions to test my current level of understanding" to allow the AI to determine how much you know about personal finance.

If you want a more dedicated teacher, use the prompt engineering coach from earlier chapters. As you write the prompt, think about what a typical finance course would look like. For some inspiration, check out the modules in online courses offered by credible education platforms like Coursera, EdX, or Udemy. Ensure it's a course on personal finance and note the topics covered in the course modules. Account for all of these in your prompt for the best outcome. Don't forget to set the tone and difficulty level, which will affect the teaching style, and whether you want to take end-of-session tests or an exam. The AI can perform all of these tasks and more.

Tip: You can also use the principles you've learned about prompting to create a more sophisticated personal coach with your preferred tone and personality.

Remember, you can always rely on AI itself to guide your prompting process. For example, I told the prompt engineering coach about the criteria to consider for writing my finance tutor prompt:

Prompt:

I'm thinking of creating a prompt that would serve as my finance tutor. The aim of the finance tutor should be to help me understand concepts related to personal finance. It should be able to create a course, develop a course outline, follow the course outline, test my knowledge, and score my performance. What parameters do you think I should specify in the prompt to achieve this?

The AI responded with a list of factors I could consider, such as:

- **Scope (defining what it covers and doesn't cover):** This included things like goals, topics, and exclusions
- **Course structure:** It listed format, time frame, and order of topics.
- **Teaching style:** Did I want a friendly or textbook-style tutor? The AI also noted methodologies, engagement, and other customizations, like knowing how to spot when I'm confused.
- **Assessment and feedback:** It asked if I wanted quizzes, knowledge checks, and a score system.

When you follow an approach where you rely on the AI to guide your prompting, you can be sure of writing a better prompt.

Another good thing about setting up a personal finance coach is the ability to use it on the go. Talk to it like you would a teacher. After all, it *is* your teacher.

Once you've gotten some basic understanding of the core concepts of personal finance, it's time to put that knowledge into practice in the task below:

Task: Following the prompting principles you've learned so far, build a financial assistant AI that will help you in managing your finances.

The prompt should meet the following criteria:

- It should be capable of using recognized frameworks, such as the three broad financial health categories, the financial health assessment score, and the four dimensions of the financial well-being framework.
- It must be able to assess your financial well-being using questionnaires and financial data, and to categorize your financial health.
- The tone and communication style should be set to your preference.
- It needs to have the ability to provide charts that visually represent spending patterns.
- It must be able to help with budgeting.
- It should offer tailored advice on improving financial well-being based on its assessment.
- It needs to be able to help set financial goals using goal frameworks.

You'll use this prompt with the follow-up prompts discussed next. Good luck!

Creating a Budget Plan

Once your learning is complete, we enter a new domain; for this, you need another chat. Here, you want to know two things: how much money is coming in, and how much money is going out (in other words, the amount you earn and the amount you spend, usually per month).

With AI as your wingman, there's no need to spend hours with calculators and spreadsheets. Your AI assistant can handle all that workload within seconds.

There are different types of budget plans you can work with, but in this lesson, we'll cover two (the 50/30/20 method and the zero-based budgeting method) just to illustrate how they can be used.

The 50/30/20 Method With AI

If you prefer the popular 50/30/20 rule (50% needs, 30% wants, 20% savings), prompt your AI like this:

> **Follow-up prompt:**
>
> Based on my income statement, create a 50/30/20 budget breakdown. Show me specific dollar amounts for needs, wants, and savings categories. Let's think step-by-step.

If you haven't submitted your income and expenses, you could include this information in the prompt:

> **Follow-up prompt:**
>
> I earn $4,500 monthly after taxes. Create a 50/30/20 budget breakdown. Show me specific dollar amounts for needs, wants, and savings categories. Include examples of which expenses fit in each category. Let's think step-by-step.

Zero-Based Budgeting With AI

If you've provided enough data on your monthly income and expenses, you could use the zero-based budgeting method, which allocates every dollar to a purpose. It's crucial that you know your monthly expenses beforehand:

> **Follow-up prompt:**
>
> Given my current expenses, can you help me create a zero-based budget? Let's think step-by-step.

If you haven't submitted your income and expenses details, then you could include them in the prompt:

Follow-up prompt:

My monthly income is $3,800. I want to allocate money for rent ($1,200), groceries ($400), car payment ($350), insurance ($200), and utilities ($150), and build an emergency fund. Assign every remaining dollar to specific categories until I reach zero. Let's think step-by-step.

Notice the instruction "Let's think step-by-step"? It's called the *chain of thought* (*CoT*) prompting method, and it's a technique we often use to help an AI outline its thinking process when solving complex problems and minimize the chances of making errors.

The thing is, AIs don't know how to count. Test this by asking an AI to count the number of "r"s in strawberry and watch it muster up different answers. When you use the CoT method, you force the AI to apply more logic in its reasoning.

Pro tip: The CoT technique can also be applied in every other use case we've covered in this book.

Insights on Expenses

AI can help you see how you spend money, find patterns in your spending, and spot any disparities in your finances. For this to work, you need to feed it the raw material of your financial life. For now, you can start simply. Upload the following documents if you have them (spanning at least one year):

- income statements
- bank statements
- credit card bills
- list of cash transactions
- spreadsheet containing expenses

If you're not ready to hand over your files, you can describe your expenses in plain language—for example, "$200 a month on groceries, $80 on streaming services, $150 on eating out, and $600 on rent."

Once that's done, ask the AI to analyze your financial well-being using your preferred framework:

Follow-up prompt:

Can you assess my financial well-being using the four dimensions of the financial well-being framework?

Response:

Overall Well-Being Snapshot

- Present security: Fragile—breaking even but not "safe."
- Present freedom: No buffer—vulnerable to shocks.
- Future security: Weak—debt repayment dominates, no savings for retirement.
- Future freedom: Limited—long-term flexibility blocked by debt and lack of savings.

As you can see, based on the dummy finance documents I provided, the AI realized that I have too many red flags.

You can follow up with more prompts asking for advice on how to remedy the situation:

Follow-up prompt:

What can I do to fix this?

Response:

Absolutely—here's the plan boiled down into four clear steps, each tied to your financial well-being dimensions:

- Step 1: Stabilize the present. Right now, you're breaking even with no cushion. Trim at least $200–$300 from flexible spending categories like "miscellaneous," dining, clothing, and entertainment.
- Step 2: Build a shock absorber. Your biggest vulnerability is having no emergency fund. Use the savings you freed up to build a starter cushion of $500–$1,000 within 3–4 months. ...

If you already have the specific advice you need from your financial assistant, you can use other prompts like these:

- What do my spending habits say about me?
- Are there any unusual spikes in my expenses this month?
- If I cut 10% from discretionary spending, where would it make the biggest difference?
- Am I overspending compared to typical budgeting rules of thumb?

That's the beauty of AI: These simple problems can be easily resolved without having to spend $150 or more per hour with a financial expert.

Financial Goal Setting

You learned about goal-setting frameworks earlier, but let me show you how to turbocharge the SMART method with AI to create financial goals that actually stick.

Before you even think about talking to an AI, get crystal clear on what you really want. Not what you think you should want, not what impresses your neighbors, but what you need to improve your life. Start the conversation with your AI like this, replacing the text in square brackets with your actual information:

> **Prompt:**
>
> I want to set a financial goal. Help me make it SMART. My broad goal is: [insert your goal here]. I currently earn [your income], I have [savings amount] saved, and my monthly expenses are approximately [amount]. What questions should I ask myself to make this goal specific and realistic?

The AI will probe deeper than most humans would dare. It'll ask about your timeline, your risk tolerance, any competing priorities, and whether this goal aligns with your broader financial picture.

Once you've defined your raw goal, work with AI to structure it using the SMART criteria:

> **Follow-up prompt:**
>
> Help me structure my goal using SMART criteria. My goal is [your refined goal]. Break this down into specific, measurable, achievable, relevant, and time-bound components. Also, identify potential obstacles I might face and suggest contingency plans.

You'll marvel at what AI is capable of when you start working on action plans and milestones. You can follow up with a prompt asking it to break your goals into manageable steps:

> **Follow-up prompt:**
>
> Based on my SMART goal of [insert your refined goal], create a detailed action plan with monthly milestones, specific steps I need to take each quarter, and key metrics I should track. Also, suggest ways to automate as much of this process as possible

AI will give you a roadmap that's far more detailed than anything most financial advisors provide in their first meeting. I'm talking specific savings amounts, investment allocation suggestions, timeline checkpoints, and even behavioral nudges to keep you on track.

"What If..." Scenario Analysis

Let's say Gerald is looking to take out a loan, but he's conflicted between two options. He can take out either a short-term loan with higher monthly payments or a long-term loan with lower payments. Each loan has its cons and pros. And that's what Gerald's struggling with: He doesn't know which will be better.

In this scenario, he can turn to AI. He can simply provide the terms of each loan and his financial situation to his AI personal finance coach and understand how each decision truly impacts his finances:

> **Prompt:**
>
> **I'm planning to take out a $20,000 loan. Should I pick a 3-year plan with a repayment of $650/month or a 5-year plan with $400/month? Can you help me understand how each choice would affect me? I earn $3,000 per month. My expenses per month sum up to $2,500 per month. Think step-by-step.**

AI can help you in different "what if..." scenarios, which you can then use to guide your decision-making. You can either add decision-making elements into your finance prompts or use another chat that focuses on decision-making. But I advise that you keep different subject matters in their own chats.

The tools are in your hands. The frameworks are proven. The only variable left is your willingness to start. Once you do, bear in mind that AI can help you in another equally important aspect of life: your health. That'll be our topic for the next chapter.

Chapter 9

Health and Wellness Advisor

ChatGPT diagnosed a woman with thyroid cancer. Let me tell you about Lauren Bannon, a 40-year-old marketing entrepreneur and mother of two who spent months watching her health deteriorate while doctors missed what was killing her. It started innocently enough, with just stiff fingers in the mornings and evenings in early 2024. When she visited her doctor, they diagnosed her with rheumatoid arthritis, despite the fact that she tested negative for the condition. But her illness got worse.

Desperate for answers, she turned to ChatGPT, asking for other medical conditions that could be confused for arthritis. Fortunately, AI mentioned Hashimoto's disease: a thyroid problem. With her new insight, she went back to doctors for more tests.

In January 2025, Lauren underwent surgery to remove her thyroid and two lymph nodes: The doctors had discovered two cancerous tumors growing in and around her thyroid ("Top Doctors Failed for 18 Months," 2025). Just think for a moment what would have happened if she hadn't asked ChatGPT. She may have continued taking the rheumatoid arthritis medication, and the cancer would've spread from her neck to everywhere else.

This raises a profound question: Can you rely on ChatGPT to assess your health? This chapter will by no means tell you to stop attending your medical appointments. Rather, you'll become more informed and learn how to support your health.

Can AI Diagnose My Condition?

Lauren Bannon's story is remarkable, but it also highlights some uncomfortable issues. How did an AI chatbot succeed where seasoned medical professionals failed for 18 months? When Lauren fed her symptoms into ChatGPT, the AI considered thousands of possible connections simultaneously. It didn't dismiss her thyroid concerns because her initial presentation looked like arthritis. It processed her complete symptom profile—the joint stiffness, the unexplained weight loss, the stomach pain—and flagged patterns that human doctors, focused on individual symptoms, had missed.

Sounds like good news—until you hear other stories of AI misdiagnosing people. A woman shared on an anonymous platform that ChatGPT misled her. She explained that she'd been battling with severe headaches for a few days and was too frustrated to wait for her doctor's appointment. So, she reached out to ChatGPT, describing her symptoms.

Can you guess what the chatbot suggested? It was two words: "brain tumor." She panicked and rushed to the emergency room, convinced she was dying. Did the scans reveal any brain tumor? No. Apparently, her migraines were triggered by stress and dehydration. Now, she had another "headache" to deal with: the $3,000 ER bill ("Top Doctors Failed for 18 Months," 2025).

Why am I sharing this with you?

Although I've showered AI with praise, it can be a double-edged sword when pushed beyond its capabilities.

AI cannot replace your doctor. It can't perform a physical exam, factor in your medical history, or understand your lifestyle factors—and, most importantly, it lacks the clinical judgment honed from years of treating real patients. AI sees patterns in data, but it can't feel a lump, hear an irregular heartbeat, or notice the subtle signs that experienced physicians pick up during face-to-face interactions.

Yet dismissing AI entirely would be throwing away a powerful tool. The key is using it intelligently: not as a replacement for medical professionals, but as a way to become a more informed patient who asks better questions and catches potential red flags.

Interpreting Test Results

Not every hospital or diagnostic center follows the same playbook. What one lab calls "normal," another may call "borderline." For example, one lab may say your fasting glucose is normal up to 100 mg/dl, while another cuts it off at 99. Same blood, different verdict.

It's not just numbers, either. Some labs use different units (mg/dl vs. mmol/l). Others adjust ranges depending on your age, sex, or pregnancy status, as a thyroid level that looks fine for a 30-year-old man could be worrisome for a pregnant woman. AI tools don't always "see" this kind of detail unless you spell it out for them, which means borderline results, like slightly high cholesterol, might get misinterpreted as a crisis.

Therefore, you'll have to feed the AI with this context. Then comes the other challenge: How do you know what information to provide if you aren't a medic yourself? The answer is straightforward: You make the AI ask for the right data. For example, you could upload your report document (or a high-resolution photo of it) into your chatbot with a prompt like this:

Prompt:

I'd like you to help me interpret my [lab results]. First, check whether the results I provide include enough context such as reference ranges, measurement units, my age, sex, and pregnancy status (if relevant), and whether multiple past results are listed for comparison. If any of these details are missing, ask me step-by-step to provide them. Once you have all the necessary information, explain my results in plain language, highlight which values are in or out of range according to the reference ranges given, and suggest some questions I should ask my doctor.

The "ask for context" clause in the prompt forces the AI to scan the report for available contexts like ranges, units, and demographics, and then ask for the context it deems missing. Then, it will proceed to interpret the results.

If you need more clarity after the AI has interpreted your results, you can always follow up with more simple prompts. For example:

Follow-up prompt:

Does this result mean I have [diabetes, anemia, thyroid disease, etc.]?

Follow-up prompt:

What do these values mean in combination?

Follow-up prompt:

How serious is this—should I be worried right now?

Follow-up prompt:

What changes can I make (diet, exercise, supplements) to improve these results?

Follow-up prompt:

Do my results vary depending on the time of day, fasting vs. eating, stress, or hydration?

Follow-up prompt:

What does this result mean for my day-to-day life?

Fitness Coaching

Fitness is the new frenzy. Like me, many people have downloaded fitness apps and found them almost useless. These apps often come with different kinds of workouts, timers, and records, which are great for motivation. However, my one complaint is the inability of the apps to customize the workouts to suit my needs. For example, one app I downloaded had the beginner regime set to 20 push-ups. Who said I could do 20 repetitions? I couldn't, and most people can't.

AI continues to give us more reasons to love it. While it may not be able to count my reps or provide an animation demonstrating the workout, it can craft workout regimes that suit your lifestyle, body type, and workout goals.

You can begin with a simple prompt such as this (notice the "Ask me questions..." clause?):

> **Prompt:**
>
> I want you to act as my fitness coach. Ask me questions to assess my current activity level, goals, strengths, limitations, and any health concerns. Then create a safe, progressive routine based on my answers

This is, in most cases enough to get things going.

Suppose you're in your 40s and already have a workout goal in mind. You could simply ask follow-up prompts aligned to that goal:

Follow-up prompt:

I'm 40, new to exercise, and want to lose weight. Please build me a 3-days-per-week workout that combines light cardio and beginner bodyweight strength training.

Follow-up prompt:

I've been lifting for 2 years. Please design a 4-day strength split focused on hypertrophy, with progressive overload.

Follow-up prompt:

I'm training for a half-marathon. Help me create a 12-week program that balances running volume with strength and recovery.

You'll find that, unlike the workout app, the AI can easily adapt to your situation without charging you extra!

Task 1: Create a Workout Coach

If you want to create a workout coach, you can use the prompt engineering coach from earlier chapters to do so. Just ensure you've included a clause that asks for more context, or include the context in the prompt. Some key parameters to provide in the prompt include:

- basic stats like age, sex, height, and weight
- activity level, whether sedentary, lightly active, moderate, or very active
- goals—for example, fat loss, muscle gain, strength, mobility, or endurance
- limitations like injuries, health conditions, and the equipment availability
- preferences—for example, workouts at home versus at the gym, or short versus long sessions

This is all context the AI will need to provide support that's close to human coach level. In addition, you'll need a framework on which the AI can base its advice.

The FITT Principle

As I said, AI doesn't know your body until you give it the information it needs. The FITT principle gives you a way to prime your AI coach with the same blueprint professional trainers use when designing programs that cost thousands.

The FITT principle represents frequency (how often you train), intensity (how hard you push yourself), time (how long each session lasts), and type (the kind of exercise you do). It can be incorporated into very simple prompts, like these:

- "I can work out three days a week." (frequency)
- "I get winded quickly, so keep it light at first." (intensity)
- "I have 30–40 minutes per session." (time)
- "I prefer bodyweight routines at home, no gym equipment." (type)

> **Prompt:**
>
> I want you to act as my fitness coach. Please design a workout plan using the FITT principle. Here are my details: [age, sex, fitness level]. I can exercise [X days] per week for [Y minutes]. I prefer [workout type]. Please create a structured routine that balances intensity and recovery, includes progression over time, and adapts to my feedback.

Task 2: Incorporate the FITT Model Into Your Health Advisor Prompt

How do you know the AI won't just throw random exercises your way? We know trainers follow a rhythm: prepare, train, recover. You can nudge the AI into doing this, too.

For example, you could include a prompt like this:

Prompt:

Build a program that follows this structure: warm-up > 2–3 rounds of core exercises (primary, secondary, and tertiary movements) > cool down. Make sure each week builds on the last with progressive overload.

As you go through each rep, set, week, or month, you can continue to refine your regime. For example, say you've completed week one. You could follow-up with: "I've completed week one. Please make week two slightly harder by increasing the reps or sets by about 10%."

What if the exercises were too tough? You could adjust by giving the feedback in the same way you would a trainer with something like: "The push-ups were too tough. Can you suggest an easier variation?"

Say you're not familiar with many exercises. In this case, you have two options: Watch a YouTube video, or ask the AI to describe the actions with a simple prompt like "Explain proper squat technique in simple steps for a beginner."

The ability to go back and forth with AI is what separates it from the regular workout apps you'll find in an app store.

Nutrition Coach

In an ABC news article, "Can ChatGPT make dinner easier? I put AI to the culinary test," the author, Kelly McCarthy, was able to create a new recipe using ChatGPT (McCarthy, 2023). She began by priming the chatbot with a list of ingredients: mango, chili paste, shrimp, lentils, and white miso. Before typing in the ingredients she had in her fridge, she asked the AI if it could "test specific fusion flavors." Then, she followed this with a simple prompt: "Can you please create a dinner recipe using frozen mango, chili paste, shrimp, lentils and white miso?"

She continued to iterate, asking ChatGPT to formulate different recipes after adding new ingredients or to create alternatives of the first recipe. By the end, Kelly, with the help of ChatGPT, came up with a recipe that she cooked at home and enjoyed: mango shrimp and lentil bowls with chili miso sauce.

Making the Nutrition Coach

Exercise may build the engine, but nutrition fuels it. Both of them are crucial for maintaining good health. If you've ever tried to count calories or figure out how much protein is in your lunch, you know it can be tedious. AI can ease the pain. Simply type in what you've eaten (or plan to eat), and prompt it as follows:

> **Prompt:**
>
> Estimate the nutrition breakdown (calories, protein, carbs, fat, fiber, sodium) of this meal: [list ingredients and portions].

The chatbot won't be perfect, but it will give you a ballpark estimate based on standard food databases. That's often enough to see whether you're hitting your daily targets or overshooting them.

Aligning Meals With Your Goals

Maybe your doctor told you to cut back on sodium, or you're aiming for more protein to support strength training. AI can help you compare your current meals with your targets:

> **Prompt:**
>
> I'm aiming for 120 grams of protein per day. Here's what I ate today: [list]. How much more do I need to hit my goal?
>
> **Prompt:**
>
> This stir-fry has 1,200 mg of sodium. Suggest three simple tweaks to lower it below 800 mg without losing flavor.

Tip: There are some apps powered by AI that can help you count the number of calories based on pictures of your meal. Use one of those along with your chatbot. You can screenshot text from the app into your chatbot.

We know nutrition isn't a one-size-fits-all. A human nutritionist would account for these variables when designing a meal. So can the AI, but you must provide the context. For example:

> **Prompt:**
>
> I'm allergic to peanuts. Can you modify this Thai recipe [input recipe] to be nut-free?
>
> **Prompt:**
>
> Give me a gluten-free alternative for chapati that still fits in a South Asian diet.
>
> **Prompt:**
>
> Make this recipe healthier. Suggest alternative ingredients, reducing calories/fat/sodium/sugar] but keeping flavor.
>
> **Prompt:**
>
> Suggest a healthier cooking method for fried chicken that keeps it crispy but lower in fat.

What if you're vegan and looking for a protein-rich diet? All you have to do is ask:

> **Prompt:**
>
> Suggest five high-protein vegan options under 400 calories.

The use cases of an AI nutrition coach don't end there. It can equally help with your workout regime. After all, food greatly impacts your body weight gain or loss goals. So, why not create an agent you can use to ensure your eating habits aren't disrupting your workout goals? For example:

Prompt:

Role: You are my fitness and nutrition sync coach, Claire.

Goal: Help me align my eating habits with my workout goals (weight loss, muscle gain, or maintenance) so that nutrition supports, not disrupts, my training progress.

Tasks:

- Start by asking me for key details:
    - My fitness goal (lose fat, build muscle, maintain)
    - My current workout routine (type, frequency, intensity)
    - My usual eating habits (meals, snacks, timing, diet style)
    - Any restrictions, preferences, or challenges (budget, time, cravings)
- Analyze this information and point out where my diet may support or conflict with my workouts.
- Suggest specific, evidence-based adjustments (meals, macros, timing, or habits).
- Provide short action steps and sample meal/exercise tweaks.
- Track my progress over time, adjust as needed, and motivate me along the way.

Style: Friendly, motivating, and knowledgeable. Use simple explanations, real-world examples, and short action steps I can follow right away.

Notice a new element you added to your prompt this time around? Your AI can have a name: Claire!

By now, you've created different bots, which may be scattered all over the place. Since some aspects of life are interconnected, you'll want one coach to be able to get information from another. For example, how do you get your nutrition coach to work with your workout coach? This will be our discourse in the next chapter.

Chapter 10

One for All or All for One?

Let's bring back the four main AI chatbots: ChatGPT, Claude, Gemini, and Copilot. Each of these costs roughly $20 per month and, as I discussed in Chapter 1, each has its perks and cons. One chatbot may perform better at certain tasks than others, which might tempt you to use ChatGPT for some things and Claude for others. But do the math for a moment. That's $40 per month, or $240 for 6 months. Half of that could cover your electricity or water bill.

This defeats the entire purpose of using AI to make your life simpler and more efficient. Hence, this is the point where you ask yourself: *Can I afford to subscribe to more than two platforms, or do I converge my AI assistants into one bot?*

You need to find a balance. And that's exactly my goal for this chapter: to improve your life on many fronts while saving on costs. So, whether you choose to consolidate everything into one powerful assistant or maintain a carefully curated team of specialists, you'll finish this chapter with a clear roadmap that serves your goals, not your AI subscriptions

Consolidating Your AI Assistants

Before you can optimize your AI strategy, you need to take an honest inventory of what you've built. A rookie mistake is opening hundreds of chats for the same purpose. That won't serve you well in the long run. And this is life we're talking about here: You want to be intentional and strategic.

So, step one is to list every AI coach you've created. For each one, clarify its unique role, its area of expertise, the platform you've created it in, and why you've chosen that platform using the framework I've provided below:

AI assistant	**Primary function**	**Platform**	**Frequency of use**	**Unique value of chosen platform**	**Cost of platform**
Savings coach	Helps me save money	Claude	Weekly	Easy integration with Google spreadsheet	$20/month

After filling in the details, you may find that you have various assistants spread across different platforms. Now, you want to ask the tough questions. Begin with these:

- Why have you chosen a specific platform for a particular AI?
- Is it the way it mimics human tone and voice? Is it the voice mode?

- Could it be that it's a cooler-looking platform? Or that you're already using it for work?
- Has the AI already been customized (via customization settings) for your preferred tone, style, and creative process?
- Do you rely on it because of its deep research or search engine access capabilities?
- Could these AIs be unified under one platform? If not, exactly why not?

The right answer depends on your workflow and communication preferences. Take into account your usage patterns from the audit above. If you're consistently using one AI chatbot for 80% of your needs, then using that same chatbot for all of your assistance probably makes sense. Consolidating your AIs works best in the following circumstances:

- You value simplicity and seamless workflow over specialized expertise.
- Your needs are broad but not extraordinarily deep in any one area.
- You want to minimize subscription costs.
- You prefer building one comprehensive relationship rather than managing multiple AI personalities.

But if you genuinely rely on specialized knowledge from different AIs, the complexity might be justified. For instance, I prefer to use Claude for creative writing because it's more inventive and more closely mimics human style than other chatbots. However, one downside of choosing Claude is that I miss out on image and video generation, which are great tools to have in this Gen Z era. I also love ChatGPT because I can generate videos with Sora, images with DALL-E, and documents with canvas, and also receive feedback through its different voice mode personalities. Thus, I keep both subscriptions.

An X (formerly Twitter) user might prefer to kill two birds with one stone by subscribing to Grok AI, so they also get a premium subscription to X. Someone looking to use multiple AIs might purchase a Poe AI subscription so they can use ChatGPT, Claude, Grok, Gemini, and Copilot together—all for $20.

Choosing a Lifelong Companion

If you're still on the fence, this worksheet should help you figure out which AI best suits your needs.

Tick the features that you often use in your AI chatbot. Next, count the number of ticks you have per chatbot and write the total in the "Total" row of the table below:

Feature	ChatGPT	Claude	Gemini	Copilot
Conversational AI	Yes ☐	Yes ☐	Yes ☐	Yes ☐
Multi-turn dialogue	Yes ☐	Yes ☐	Yes ☐	Yes ☐
Text input	Yes ☐	Yes ☐	Yes ☐	Yes ☐
Audio input	Yes ☐	No ☐	Yes ☐	Limited ☐
Voice input	Yes ☐	No ☐	Beta stage ☐	Limited ☐
Voice output (TTS)	Yes ☐	No ☐	Beta stage ☐	Limited ☐
Image input	Limited ☐	No ☐	Yes ☐	None ☐
Image generation	Yes ☐	No ☐	Yes ☐	No ☐
Video input	No ☐	No ☐	Yes ☐	No ☐
Video generation	No ☐	No ☐	Yes ☐	No ☐
Code generation	Yes ☐	Yes ☐	Yes ☐	Yes ☐
Code debugging	Yes ☐	No ☐	Yes ☐	Yes ☐
Language translation	Yes ☐	Yes ☐	Yes ☐	Yes ☐

Summarization	Yes ☐	Yes ☐	Yes ☐	Yes ☐
Real-time web search	Yes ☐	No ☐	Yes ☐	Yes ☐
Plug-ins/integrations	Extensive ☐	API/partners ☐	Google ecosystem ☐	Microsoft 365 integration ☐
Workflow integration	Growing ☐	Limited ☐	Google Workspace ☐	Strong ☐
File upload	Yes ☐	Limited ☐	Yes ☐	Yes ☐
Document analysis	Yes ☐	Yes ☐	Yes ☐	Yes ☐
Project/task management	Yes ☐	Limited ☐	Yes ☐	Yes ☐
Personalization/memory	Yes ☐	No ☐	Yes ☐	Yes ☐
Safety focus	Moderate ☐	Strong ☐	Moderate ☐	Enterprise focused ☐
Large context window	Yes ☐	Yes ☐	Yes ☐	Limited ☐
Notifications	Yes ☐	No ☐	No ☐	Yes ☐
Enterprise security	Moderate ☐	Moderate ☐	Moderate ☐	Strong ☐
Collaboration tools	Canvas-based ☐	No ☐	Google Docs ☐	Teams integration ☐
Voice typing	No ☐	No ☐	Experimental ☐	Yes ☐
API access	Public ☐	Selective ☐	Limited ☐	Enterprise ☐
Total				

AI can help you make the final decision, too. Let's write a prompt that helps you figure out which AI features you use the most for productive tasks. I call it the "Bot Matchmaker":

Prompt:

Role: You are my Bot Matchmaker.

Goal: Help me figure out which AI bot (ChatGPT, Claude, Gemini, Grok, Copilot, Poe, among others) best matches my needs.

Tasks:

1. Ask me one playful, relatable question at a time about how I want to use AI.
   - Example: Instead of "Do you generate videos?" ask, "Do you think you'll want to whip up a fun video to post online or share with friends?"
2. Wait for my answer before moving to the next question.
3. Use my answers to build a profile of what I need in an AI bot (content creation, productivity, research, career help, finances, fitness, etc.).
4. Recommend the AI bot that covers the highest number of my needs, and explain why in simple, clear language.
5. Give me optional alternatives if another bot is close to what I need.

Style: Conversational, fun, and friendly, like a buddy helping me pick a new app. Keep questions light and engaging, not technical or overwhelming.

In the end, whatever you choose, what matters most is effectiveness: Your AI setup should serve toward reaching for your goals.

Reminder!

This is just a friendly reminder to not have your agents sounding the same. I showed you how to give personalities to your AI bots in Chapter 5. Review the nutrition, fitness, finance, career, and emotional coaches you've created. Adopt any of the prompting frameworks covered in Chapters 4–8 to give each of these bots its unique personality. Let each one be a different "person." Play around with the settings. The more you do, the better you'll become, and the more fun your AI will be!

Is AI Stealing Your Data?

The internet was set ablaze when word got out that AI companies were training their chatbots using content generated by humans. Large media companies like *The New York Times* and Getty Images were some of the first to throw punches. These companies sued AI juggernauts like Anthropic, Meta, Google, and OpenAI, accusing them of copyright infringement. But large companies aren't the only ones worried about AI "stealing" their data. Everyday folks have become part of the mix.

Many of my clients raise concerns when I advise them to feed AI with enough context. Some ask, "Is it safe to upload documents?" Others worry their data may be sold to third parties. These are all valid concerns—ones you should also take seriously. People do have the right to be worried. After all, we saw the news about AI corporations using people's content to train AI without asking for consent from the creators.

Will AI companies steal your data? The simple answer is no. The legal answer isn't so straightforward. While AIs can't steal your data, they do have access to it, and AI companies do confess to using your data to train their models.

The undeniable truth is that AI systems often collect and use data in ways we don't fully understand or meaningfully consent to. OpenAI clearly states that it uses your conversations to improve its models, unless you opt out of it (*How we're responding to The New York Times,* 2025). Although Anthropic (Claude) maintains that it respects user privacy, it still collects usage data for safety reasons. Google's Gemini connects with your broader Google ecosystem. In other words, it can see your schedules and look through the files you have stored in your Google Drive. That's a lot of data! It's the same for Copilot, which links up with your Office 365 data.

In short, AI has access to your data. But this shouldn't cause any alarm. Governments have put laws in place to restrict AI companies from illegally using your data. For instance, back in 2024, the European Union rolled out tough new rules with its AI Act (Regulation 2024/1689, 2024)—one comprehensive regulatory framework that makes sure AI systems are safe, transparent, and nondiscriminatory, and that they respect users' privacy.

Then again, laws aren't set in stone. I mean, who knows—privacy policies might have changed by the time you get to this page. Your data might be protected today but not so secure tomorrow. Either way, you bear some responsibility for keeping your data safe.

Protecting Your Data

Should you abandon AI? Well, you don't need to. That said, you should be strategic about your data exposure. You can take the following measures:

- Turn off the conversation history where possible. Most platforms now offer this option—use it for sensitive discussions.
- You could create dedicated email accounts for AI services, separate from your primary personal or work accounts.
- Try anonymizing sensitive information. For example, instead of "I'm considering divorce..." try "A character called Adam is divorcing..."
- Always turn on two-factor authentication security, and use strong, not easily predictable passwords. Never use personal details as your password!
- Think twice before you share any documents. You have the authority to share your own information, but it is illegal to share that of others. In addition, never share your office work with AI, unless you're allowed to do so.

Should you allow AI to use your data? That's totally up to you. Do bear in mind that AI companies need your data to improve their services. So, on one hand, you're doing the world a favor. On the other hand, you're exposing your precious data.

Having said that, I'll tell you this: Make sure you're getting equivalent value in return, and that you're comfortable with the trade-offs you're making. Trust, but verify. And when in doubt, keep your most sensitive information offline.

We've finally reached the last lap of this awesome marathon, and I have one last question for you: What will you build with these new powers of yours?

Conclusion

Congratulations! You've not only supercharged your AI skills, but you've become a more productive individual and also completed a journey that most people haven't even begun. Frankly speaking, many are still too intimidated or skeptical to start.

Take a moment here. Really pause and recognize the transformation you've achieved. When you flipped to page 1 of this book, *Prompted by Feelings*, AI was another fun toy. But look where you are now.

You've learned to turn a simple chatbot into an emotional intelligence coach—one that probes into your deepest feelings and sense of being. This adventure continued as you created a financial advisor capable of exposing your secret spending habits, helping you lay out SMART goals, and even interpreting bank statements.

Through Chapters 3 to 9, you learned to prompt AI following best practices and created AI personas with unique communication styles and duties. Combined, these personas will serve to turbocharge your growth—whether in your career, relationships, health, or finances—based on the four pillars of growth framework.

The Four Pillars of Growth

Without a doubt, this book has shifted your mindset and added prompt engineering to your skill set. Life, in the long run, should get better as you continue to evolve across the four foundational pillars we've built together:

- **Emotional intelligence:** Very few people are aware of the emotional continuum through which they experience life. CBT and DBT are tools that are unfamiliar to most. But with the help of AI, you've moved from simply feeling your emotions to naming and understanding them. In addition, you've learned how to track your emotional patterns and developed better self-awareness and emotional regulations strategies.
- **Decision-making skills:** Making decisions takes a great deal of brain power. It involves thinking, weighing risk, and making the right choice. You've learned to lessen this burden by seeking advice from your decision-making AI assistants, which make use of frameworks to gather information, create new ideas, and solve problems.
- **Productivity optimization:** You've transformed from someone who struggled to set goals, or didn't quite understand the mechanics of doing so, to someone with a range of project management and goal-setting frameworks under their belt. All thanks to the knowledge you've gained about instructing chatbots to use expert-level project management systems.

- **Social confidence:** Humans are social creatures, and we depend on our connections with others to survive. Through the productive use of role-playing with AI, you've developed skills that will boost your confidence at social events, during job interviews, in the workplace, and so on. Maybe a month ago, words struggled to leave your lips. Now, they flow smoothly and gain you recognition.

As you struggle through the storms of life, don't measure your growth by comparing yourself to others or by some external standard of success. A mentor once told me, "Only compare yesterday's you with today's you. Then, look forward to making tomorrow's you better than yesterday's you."

So, measure your growth by the skills you've gained through working with AI. Are you a better communicator than you were one week ago? Do you brainstorm and solve problems better than you did a month ago? Which emotional expressions do you need some more time to master? You'll find that life is more satisfying when it blossoms at its own pace, not based on the time stamps set by others.

Spread the Word

As you become a better version of yourself, share your wisdom with others, just as I've shared mine with you. Look around at your friends, family members, and colleagues. Are any of them struggling with the same problems this book has helped you solve? How many of them can afford to hire a career coach or life coach?

Do you know what? You now have the tools to help them—and, more importantly, to teach them how to help themselves. Share what you've learned. Ask them to pick up this book. Show them how AI can be a supportive coach rather than an inhuman replacement. Let them know that the future belongs to people who stay flexible and embrace new tools without getting stuck in conservative paralysis.

Finally, don't close this book and abandon it on the shelf. Keep experimenting. Keep learning. And keep sharing what you discover. I would love to hear how this book has inspired and shaped your life. Please leave a review to show the world that AI can help them turn things around!

Let's go!

Congratulations — You Did It!

You've reached the end of Prompted by Feelings. That alone is an accomplishment worth celebrating. You've walked through ideas, exercises, and insights designed to help you see AI in a new light — not as a machine, but as a partner in growth.

Finishing a book is like crossing a finish line: you've invested your time, your focus, and your curiosity. Now, would you be willing to invest just a few more seconds to help others find this journey, too?

A short review — even a sentence or two — helps future readers know that this book is worth their time.

Your voice matters. Your progress matters. And your words could inspire the next person to begin.

---

or

Leave a Review
Goodreads
https://www.goodreads.com/book/show/242613490

Visit Our Website
ViPR Publishing
https://viprpublishing.com/prompted-by-feelings

Thank you for reading. We hope to see you again soon.

Ancora Imparo — Always Learning

References

AI companion market size, share & trends analysis report by type (text-based, voice-based, multi-modal), by application (mental health support, education & learning aid), by industry vertical (consumer, businesses, healthcare), and by region forecasts, 2025 - 2030. (2025, February). Grandview Research. https://www.grandviewresearch.com/industry-analysis/ai-companion-market-report

Alanezi, F. (2024). Assessing the effectiveness of ChatGPT in delivering mental health support: A qualitative study. *Journal of Multidisciplinary Healthcare, 17*, 461–471. https://doi.org/10.2147/jmdh.s447368

American Society for Quality. (n.d.). Problem solving. https://asq.org/quality-resources/problem-solving

Aristotle. (n.d.). *A quote by Aristotle*. Goodreads. https://www.goodreads.com/quotes/3102-knowing-yourself-is-the-beginning-of-all-wisdom

Benefits of self-awareness: Increase emotional intelligence and wellbeing. (2024, November 21). *Life Tools Programme*. https://blogs.reading.ac.uk/life-tools-programme/benefits-of-self-awareness-increase-emotional-intelligence-and-wellbeing/

Borghesan, K. (n.d.). *Americans are struggling financially: 13 financial statistics you need to know*. Savology. https://savology.com/13-financial-statistics-you-need-to-know

Carter, P. (2025, July 18). *The AI companionship market and its explosive growth potential*. AInvest. https://www.ainvest.com/news/ai-companionship-market-explosive-growth-potential-2507/

Clear, J. (n.d.). *The mistake smart people make: Being in motion vs. taking action*. James Clear. https://jamesclear.com/taking-action

Covey, S. R. (n.d.). *Stephen R. Covey quote*. Goodreads. https://www.goodreads.com/quotes/1011157-trust-is-the-highest-form-of-human-motivation-it-brings

Deming, W. E. (2015, February 26). *A bad system will beat a good person every time*. The Deming Institute. https://deming.org/a-bad-system-will-beat-a-good-person-every-time/

Doctor Mike. (2025, April 9). Which A.I. would make the best doctor? [Video] YouTube. https://www.youtube.com/watch?v=WLqLA5XnfiE

Fountain House. (2022, August 30). *4 out of 10 Americans can't access mental health care when they need it – community-based support is an immediate solution*. https://www.fountainhouse.org/news/4-out-of-10-americans-cant-access-mental-health-care-when-they-need-it-community-based-support-is-an-immediate-solution

Freedman, J. (2024, February 29). *Dr. Daniel Goleman explains the history of emotional intelligence*. Six Seconds. https://www.6seconds.org/2024/02/29/goleman-emotional-intelligence/

Gabb, J., & Singh, R. (2014). The uses of emotion maps in research and clinical practice with families and couples: Methodological innovation and critical inquiry. *Family Process, 54*(1), 185–197. https://doi.org/10.1111/famp.12096

Greene, R. (n.d.). *Robert Greene quote.* Goodreads. https://www.goodreads.com/quotes/740113-the-future-belongs-to-those-who-learn-more-skills-and

Guirdham, O. (2025, June 13). *Key trend shaping the life coaching services market In 2025: AI-powered coaching revolutionizes personalized support.* EIN Presswire. https://www.einpresswire.com/article/821883384/key-trend-shaping-the-life-coaching-services-market-in-2025-ai-powered-coaching-revolutionizes-personalized-support

How we're responding to The New York Times' data demands in order to protect user privacy. (2025, June 5). OpenAI. https://openai.com/index/response-to-nyt-data-demands/

Lim, L. (2025, August 2). *Is ChatGPT updated to 2025? Exploring the latest AI developments.* BytePlus. https://www.byteplus.com/en/topic/547773?title=is-chatgpt-updated-to-2025

Johnson, J. (2020, July 7). *What is decision fatigue?* Medical News Today. https://www.medicalnewstoday.com/articles/decision-fatigue

Kiyosaki, R. T. (2020, April 4). *Financial freedom is a mental, emotional and educational process* [Status update] Facebook. https://www.facebook.com/RobertKiyosaki/posts/financial-freedom-is-a-mental-emotional-and-educational-process-richdad/10158160533191788/

Leahey, T. M., & Wing, R. R. (2013). A randomized controlled pilot study testing three types of health coaches for obesity treatment: Professional, peer, and mentor. *Obesity*, 21(5), 928–934. https://doi.org/10.1002/oby.20271

Lohr, S. (2023, June 26). A.I. may someday work medical miracles. For now, it helps do paperwork. *The New York Times.* https://www.nytimes.com/2023/06/26/technology/ai-health-care-documentation.html

Matthews, B., & Webster, W. (2024, December 9). *What is a RACI matrix? Definition, examples, uses.* Project Management. https://project-management.com/understanding-responsibility-assignment-matrix-raci-matrix/

May, R. (n.d.). Quote. In quoteresearch. (2018, February 18). *Between stimulus and response there is a space. In that space is our power to choose our response.* Quote Investigator. https://quoteinvestigator.com/2018/02/18/response/

McCarthy , K. (2023, July 15). *Can ChatGPT make dinner easier? I put AI to the culinary test.* ABC News. https://abcnews.go.com/GMA/Food/chatgpt-make-dinner-easier-put-culinary-test/story?id=97992302

Menefee, D. S., Ledoux, T., & Johnston, C. A. (2022). The importance of emotional regulation in mental health. *American Journal of Lifestyle Medicine, 16*(1), 28–31. https://doi.org/10.1177/15598276211049771

Mental Health Foundation. (n.d.). *People seeking help for diagnosed mental health problems: Statistics.* https://www.mentalhealth.org.uk/explore-mental-health/statistics/people-seeking-help-diagnosed-mental-health-problems-statistics

Merton, S. (2023, March 8). *"A journey of a thousand miles begins with a single step"* -

meaning and history. Stoic Quotes. https://stoicquotes.com/journey-of-a-thousand-miles/

Nartey, J. (2025, June 23). *AI job displacement analysis* (2025-2030). Social Science Research Network. https://papers.ssrn.com/sol3/papers.cfm?abstract_id=5316265

Pandey, A. (2025, July 7). *The future of AI companions – market size and growth potential*. Nasscom. https://community.nasscom.in/index.php/communities/data-science-ai-community/future-ai-companions-market-size-and-growth-potential

Pignatiello, G. A., Martin, R. J., & Hickman, R. L. (2018). Decision fatigue: A conceptual analysis. *Journal of Health Psychology*, 25(1), 123–135. https://doi.org/10.1177/1359105318763510

Regulation 2024/1689. *Regulation (EU) 2024/1689 of the European Parliament and of the Council*. (2024, July 12). https://eur-lex.europa.eu/eli/reg/2024/1689/oj/eng

Segal, J., Smith, M. A., & Robinson, L. (2025, August 15). *Improving emotional intelligence (EQ)*. Help Guide. https://www.helpguide.org/mental-health/wellbeing/emotional-intelligence-eq

Shen, Y. N. (2025, April 25). *Prompt engineering by Andrew Ng: A comprehensive guide*. BytePlus. https://www.byteplus.com/en/topic/421158

Shulman, D. (2025, August 21). *The Matrix: Film by Andy and Larry Wachowski [1999]*. Britannica. https://www.britannica.com/topic/The-Matrix

Top doctors failed for 18 months, then AI diagnosed mystery illness: Woman shares how ChatGPT "healed" her. (2025, July 24). *The Economic Times*. https://economictimes.com/magazines/panache/top-doctors-failed-for-18-months-then-ai-diagnosed-mystery-illness-woman-shares-how-chatgpt-healed-her-mom/articleshow/122885926.cms

Tzu, L. (n.d.). *A quote by Lao Tzu*. Goodreads. https://www.goodreads.com/quotes/21535-the-journey-of-a-thousand-miles-begins-with-a-single

Wiggers, K., Corrall, C., Stringer, A., & Park, K. (2025, August 29). *ChatGPT: Everything you need to know about the AI-powered chatbot*. Tech Crunch. https://techcrunch.com/2025/08/29/chatgpt-everything-to-know-about-the-ai-chatbot/

Wilde, O. (2019). *Oscar Wilde quote*. Goodreads. https://www.goodreads.com/quotes/19884-be-yourself-everyone-else-is-already-taken

Winland, V., Kavlakoglu, E., & Gadesha, V. (2024, August 12). *What is chain of thought (CoT) prompting?* IBM. https://www.ibm.com/think/topics/chain-of-thoughts

www.ingramcontent.com/pod-product-compliance
Lightning Source LLC
LaVergne TN
LVHW100318210926
848480LV00003B/1173